cosenza
& ASSOCIATES
innovation in education

ALGEBRA 1
TEKS COMPANION GUIDE
Student Edition

A Review Guide for the Algebra 1
Texas Essential Knowledge and Skills

PAUL GRAY, JR.

Algebra 1 TEKS Companion Guide

A review guide for the Algebra 1 Mathematics Texas Essential Knowledge and Skills

Student Edition

Authoring Team:

Dr. Paul Gray, Cosenza & Associates, LLC, Dallas, TX
Dr. Kelli Mallory, Cosenza & Associates, LLC, Frisco, TX

Editorial Team:

Ian Molenaar, Cosenza & Associates, LLC, Dallas, TX
Gary Cosenza, Cosenza & Associates, LLC, Dallas, TX
Judy Rice, Independent Consultant, Houston, TX

Layout Design:

Five J's Design, Aubrey, TX

© 2018 by Cosenza & Associates, LLC.

cosenza & ASSOCIATES
innovation in education

Printed in the United States of America.
Signature Book Printing, www.sbpbooks.com

ISBN 978-0-9972265-7-7

1 2 3 4 5 6 7 8   24 23 22 21 20 19 18 17

TABLE OF CONTENTS

QUADRATIC FUNCTIONS AND EQUATIONS

EXPONENTIAL FUNCTIONS AND EQUATIONS

Unit 1
Number and
Algebraic Methods

ALGEBRA 1 TEKS COMPANION GUIDE Student Edition

ADDING AND SUBTRACTING POLYNOMIALS

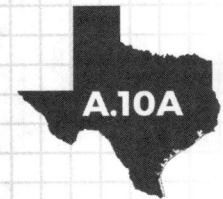

A.10A The student is expected to add and subtract polynomials of degree one and degree two.

ℹ️ TELL ME MORE...

A **polynomial** is an expression that is a sum of several terms. Polynomials may contain multiple variables, real number coefficients, and whole number exponents. Polynomials may not contain division with variables or non-whole number exponents.

Types of polynomials are named by the number of terms contained in the polynomial. **Monomials** have one term, **binomials** have two terms, and **trinomials** have three terms. Polynomials, of course, may have four or more terms as well.

The degree of a polynomial is its greatest exponent. For example, if the term in the polynomial with the greatest exponent is $7x^8$, then the polynomial is of *degree eight*. The polynomial $5.4x^2 - 8.3x + 11$ is of *degree two* since the term with the greatest exponent has an exponent of two.

Addition is an operation where two quantities are being combined. To add polynomials symbolically, combine like terms. **Subtraction** is an operation where one quantity is being removed or separated from another. To subtract polynomials symbolically, you take the opposite sign of each term in the subtrahend, or polynomial being subtracted, then combine like terms.

> **Monomials** have one term.
> Examples: $4, 4x, \frac{2}{3}x^2y^3$
>
> **Binomials** have two terms.
> Examples: $x + 4, 3b - 4c,$
> $\quad 5xy + 1.8x^2y$
>
> **Trinomials** have three terms.
> Examples: $x^2 - \frac{2}{3}x + 4,$
> $\quad 3b^4 - 4c + 89d^7$

✓ EXAMPLES

EXAMPLE 1: Simplify the expression $(6x^2 - 5xy + 10) - (15x^2 + 6.5xy - 18)$.

STEP 1 The second polynomial is being subtracted from the first. Think of the minus sign as −1 being distributed to the second polynomial and distribute the −1 to all terms of the second polynomial. Thus, you will change the signs of all terms in the second polynomial.

$$(6x^2 - 5xy + 10) - (15x^2 + 6.5xy - 18)$$
$$6x^2 - 5xy + 10 - 15x^2 - 6.5xy + 18$$

$6x^2 - 5xy + 10 - 15x^2 - 6.5xy + 18$

STEP 2 Combine like terms.

$$6x^2 - 5xy + 10 - 15x^2 - 6.5xy + 18$$
$$(6x^2 - 15x^2) + (-5xy - 6.5xy) + (10 + 18)$$
$$-9x^2 + (-11.5xy) + 28$$

$-9x^2 - 11.5xy + 28$

EXAMPLE 2: MacArthur Park has two rectangular fountains. The area of one fountain is $7y^2 - 14y + 10$ and the area of the second fountain is $3y^2 + 8y - 6$. Write an expression that represents the combined area of both fountains.

STEP 1   Determine the operation that you will need to perform to the two polynomials.

The area of the two fountains is being combined. This action can be represented with addition.

Addition

STEP 2   Add the two polynomials together by combining like terms.

$$(7y^2 - 14y + 10) + (3y^2 + 8y - 6)$$
$$(7y^2 + 3y^2) + (-14y + 8y) + (10 - 6)$$
$$10y^2 - 6y + 4$$

$10y^2 - 6y + 4$

EXAMPLE 3: The population of Washington City is represented by the polynomial $5000 + 150x$. The population of Lincolnville is represented by the polynomial $6000 + 125x$. How many more people live in Lincolnville than in Washington City?

STEP 1   Determine the operation that you will need to perform to the two polynomials.

You are looking for how much larger one quantity is than the other, which is a difference between the two quantities. Use subtraction.

Subtraction

STEP 2   Write a subtraction sentence. Subtract Washington City from Lincolnville.

$$\text{Lincolnville} - \text{Washington City}$$
$$(6000 + 125x) - (5000 + 150x)$$

$(6000 + 125x) - (5000 + 150x)$

STEP 3   Take the opposite signs of the second polynomial (subtrahend) and then combine like terms.

$$(6000 + 125x) - (5000 + 150x)$$
$$6000 + 125x - 5000 - 150x$$
$$(6000 - 5000) + (125x - 150x)$$
$$1000 + (-25x)$$

$1000 + (-25x)$

YOU TRY IT!

The number of points scored last year and this year by a basketball team is shown in the table. How many total points were scored in both years?

Points Scored

Last Year	$1.5x^2 + 17x + 1$
This Year	$2x^2 - 10.5$

Operation: _____

Polynomial Number Sentence:

_____

_____

Total: _____

NUMBER AND ALGEBRAIC METHODS   **3**

PRACTICE

Add or subtract the polynomials as indicated.

1. $(4x - 7) + (3.5x - 4.9)$

2. $\left(\frac{1}{2}x^2 + 2x - 10\right) + \left(\frac{3}{2}x^2 - \frac{2}{3}x + 9\right)$

3. $(4.8m^2 + 18mn - 19) + (7.1mn + 21)$

4. $(4.9p^2 - 10) - (6.8p^2 + 5.6)$

5. $(17.4q^2 + 7q - 9) - (11.3q^2 - 9q - 1.7)$

6. $(58x^2 - 10) - (4x + 13)$

7. Two sides of a triangle measure $3n + 2$, and $4n^2 - n + 6$. The perimeter of the triangle is $6n^2 + 5n + 8$. What is the length of the third side of the triangle?

8. Rob mows lawns for extra money. On Monday he mowed $2x^2 + 5x - 3$ yards. On Tuesday he mowed $4x - 7$ yards. On Wednesday he mowed $3x^2 + 10$ yards. How many yards did he mow over the 3-day period?

9. Kameron has $(4x + 10)$ toy cars. His best friend Joey has $(-5x + 20)$ toy cars. How many more toy cars does Kameron have than Joey has?

10. Ross and his family like to visit a pizza restaurant that also has games. On his last visit Ross collected $2t^2 + 3t + 7$ tickets playing games. After examining all the prizes he decided to save for a prize with a price of $3t^2 - 8t - 2$ tickets. How many more tickets does Ross need to earn on their next visit to be able to buy the prize?

11. Simplify
$(3m^2 - 5m) - (4 - 2m) + (1 - 3m - m^2)$

12. In 2015, the Cuppa Company earned $\frac{1}{2}x^2 + \frac{11}{4}x + 200$ dollars in revenue. In 2016 the company earned $\frac{2}{3}x^2 - \frac{7}{5}x - 548$ dollars in revenue. What was the Cuppa Company's combined revenue for 2015 through 2016?

13. The cost of Mia's electric bill for this month is $2n^2 - 5n - 17$ and the cost of her gas bill is $n^2 + 2n - 8$. Mia's cable and internet bill this month is $-4n^2 + 3n + 12$. What is the total cost of Mia's utilities this month?

14. Simplify $(-2a^2 + 7ab - 4) - (-3a^2 + 6ab - 5)$.

15. Sandra solved the following problem in class.

$$(-2x^2 + 7) - (3x^2 + x - 5)$$

Sandra's answer was $-5x^2 + x + 2$. Which of the following best explains the accuracy of Sandra's answer?

A Sandra's answer is correct.

B Sandra didn't fully distribute the subtraction. The answer is $-5x^2 - x + 2$.

C Sandra didn't fully distribute the subtraction. The answer is $-5x^2 - x + 12$.

D Sandra didn't subtract correctly. The answer is $x^2 - x + 12$.

16. Jamal's net worth is based on the sum of the value of his assets reduced by his liabilities. Jamal's assets and liabilities are shown below in dollar values.

Assets	Liabilities
Vehicle value $3x - 16$	Vehicle debt $8x + 9$
Home value $4x^2 + 2 - 12$	Mortgage debt $0.5x^2 - 4x + 3$
Total cash $x^2 - x + 25$	Credit debt $-1.5x^2 + 3x - 7$

What is Jamal's total net worth?

F $-6x^2 + 4x + 8$

G $3x^2 - 6x + 2$

H $6x^2 - 4x - 8$

J $4x^2 + 10x + 2$

MULTIPLYING POLYNOMIALS

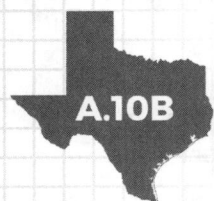

A.10B The student is expected to multiply polynomials of degree one and degree two.

ℹ TELL ME MORE...

A **polynomial** is an expression that is a sum of several terms. Polynomials may contain multiple variables, real number coefficients, and whole number exponents. Polynomials may not contain division with variables or non-whole number exponents.

Types of polynomials are named by the number of terms contained in the polynomial. **Monomials** have one term, **binomials** have two terms, and **trinomials** have three terms. Polynomials, of course, may have four or more terms as well.

The **degree** of a polynomial is its greatest exponent. For example, if the term in the polynomial with the greatest exponent is $7x^8$, then the polynomial is of *degree eight*. The polynomial $5.4x^2 - 8.3x + 11$ is of *degree two* since the term with the greatest exponent has an exponent of two.

To multiply polynomials symbolically, use the distributive property. Multiply each term in the first factor by each term in the second factor. Then, use the properties of algebra to simplify the expression representing the product.

Monomials have one term.

Examples: $4, 4x, \frac{2}{3}x^2y^3$

Binomials have two terms.

Examples: $x + 4, 3b - 4c,$
$\quad 5xy + 1.8x^2y$

Trinomials have three terms.

Examples: $x^2 - \frac{2}{3}x + 4,$
$\quad 3b^4 - 4c + 89d^7$

✓ EXAMPLES

EXAMPLE 1: Multiply $(5x - 7)(x + 2)$.

STEP 1 Use the distributive property to multiply the first term in the first factor, $5x$, by both terms in the second factor.

$$(5x - 7)(x + 2)$$
$$5x(x) + 5x(2)$$
$$5x^2 + 10x$$

$5x^2 + 10x$

STEP 2 Use the distributive property to multiply the second term in the first factor, $5x$, by both terms in the second factor.

$$(5x - 7)(x + 2)$$
$$-7(x) - 7(2)$$
$$-7x - 14$$

$-7x - 14$

STEP 3 Combine the two partial products.

$$(5x - 7)(x + 2) = [5x^2 + 10x] + [-7x - 14] = 5x^2 + 10x - 7x - 14$$
$$(5x - 7)(x + 2) = 5x^2 + 3x - 14$$

$5x^2 + 3x - 14$

EXAMPLE 2: Ian marked the boundary of a beach volleyball court as shown. The dimensions of the court are given in meters. Write an expression that represents the area of the beach volleyball court in square meters.

STEP 1 The area of a rectangle can be calculated using the formula $A = lw$. Let l and w each represent one dimension. Substitute the dimensions into the area formula.

$$A = lw$$
$$A = \left(\frac{7}{3}x\right)(6x - 21)$$

$\left(\frac{7}{3}x\right)(6x - 21)$

6x - 21

$\frac{7}{3}x$

STEP 2 Use the distributive property to multiply the first factor by both terms of the second factor.

$$\left(\frac{7}{3}x\right)(6x - 21)$$
$$\left(\frac{7}{3}x\right)(6x) + \left(\frac{7}{3}x\right)(-21)$$
$$14x^2 - 49x$$

$14x^2 - 49x$

STEP 3 Use graphing technology to check your answer. Graph $Y1 = \left(\frac{7}{3}x\right)(6x - 21)$, the product of the two dimensions, and $Y2 = 14x^2 - 49x$, the simplified product. If the graphs coincide, then the expressions are equivalent. You can also use tables to verify equivalence.

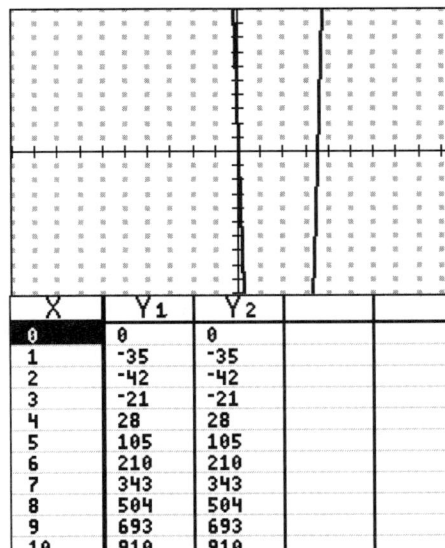

X	Y1	Y2	
0	0	0	
1	-35	-35	
2	-42	-42	
3	-21	-21	
4	28	28	
5	105	105	
6	210	210	
7	343	343	
8	504	504	
9	693	693	
10	910	910	

YOU TRY IT!

The area of a triangle is determined using the formula $A = \frac{1}{2}bh$, where b represents the base length and h represents the height of the triangle. Write an expression for the area of a triangle with a base length of $(5x - 4)$ feet and a height of $(3x + 7)$ feet.

$A = \frac{1}{2}$ (_____)(_____)

Use the distributive property:

Simplify the expression:

The graphs coincide and table shows equivalent *y*-values for each *x*-value. The expressions are equivalent

NUMBER AND ALGEBRAIC METHODS

EXAMPLE 3: For what value of k will the graphs of $y = 3x^2 - 24x + k$ and $y = 3(x - 4)^2 + 15$ be the same? Record your answer and fill in the bubbles on your answer document.

STEP 1 Rewrite $y = 3(x - 4)^2 + 15$ in standard form. Use the distributive property and other properties of algebra to simplify the expression.

$$y = 3(x - 4)^2 + 15$$
$$y = 3(x - 4)(x - 4) + 15$$
$$y = 3(x^2 - 8x + 16) + 15$$
$$y = 3x^2 - 24x + 48 + 15$$
$$y = 3x^2 - 24x + 63$$

$y = 3x^2 - 24x + 63$

STEP 2 Compare the two functions to determine the value of k.

$$y = 3x^2 - 24x + k$$
$$y = 3x^2 - 24x + 63$$

Since the x^2 term and x term are the same in both functions, the constant term should also be the same in both functions for the functions to be equivalent and their graphs the same.

$k = 63$

STEP 3 Since the question is a gridded response question, enter your response on the grid provided. Practice using the grid with the instructions.

1. Record a 6 in the first column containing numbers. Record a 3 in the next column.

2. Bubble the 6 beneath the numeral 6. Bubble the 3 beneath the numeral 3.

PRACTICE

Multiply the following pairs of polynomials.

1. $3x$ and $(4x - 5)$

2. $\frac{1}{2}x$ and $8x$

3. $1.5x$ and $(4x^2 - 10x + 20)$

4. $(3m - 7)$ and $(11m + 2)$

5. $(7b + 8)$ and $(2b^2 - 5b + 8)$

6. $(3z^2 - 6z + 10)$ and $(z^2 + 2z - 1)$

7. On a recent road trip, Patrick traveled at an average rate of $3x + 5$ miles per hour and reached his destination in $4x - 1$ hours. Write an expression in standard form that represents the distance Patrick traveled ($d = rt$).

9. The bottom of Cinnamon's dog crate has a length 6 inches greater than its width, w. Cinnamon's owner Anna plans to get Cinnamon a new crate that is 4 inches larger on both the length and width. Write an expression in standard form that represents the area, A, of Cinnamon's new crate in terms of w, the width of her current crate?

8. The diagram below shows a triangular garden with dimensions as marked.

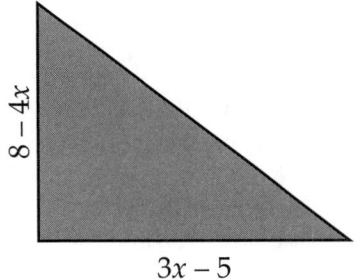

$8 - 4x$

$3x - 5$

The area of a triangle is found using the formula $A = \frac{1}{2}bh$ where A is the area, b is the base length of the triangle, and h is the height of the triangle. Write a trinomial expression to represent the area of the garden in square units.

10. The area of a trapezoid is found using the formula $A = \frac{h}{2}(b_1 + b_2)$, where A is the area of the trapezoid, h is the height, and b_1 and b_2 are the lengths of the trapezoid's bases. What expression in standard form represents the area of a trapezoid with bases $(x - 6)$ and $(3x + 4)$ inches long and a height of $5x$ inches?

11. What degree 4 expression is equivalent to $3a(4a^2 - a + 12)(7 - a)$?

12. Write an expression for the area of a rectangle with a length of $(x + 5)^2$ feet and a width of $(3 - x)$ feet.

13. The surface area of a cube can be calculated using the formula $A = 6s^2$, where s represents the edge length of the cube. If the edge length of a cube is $(2x + 7)$ centimeters, the surface area of the cube is equal to $24x^2 + 168x + c$. What is the value of c? Record your answer and fill in the bubbles on your answer document.

14. A packing box has a length of $(2x + 3)$ units, a width of $(4 - x)$ units, and a height of $(3x - 4)$ units. The volume of the box can be determined by using the formula $V = Bh$ where V is the volume, B is the area of the base of the box, and h is the height of the box. What expression can be used to find V, the volume of the box in terms of x in simplified form?

A $V = -6x - 48$

B $V = -6x^3 + 8x^2 + 36x - 48$

C $V = -6x^3 + 41x^2 - 8x - 48$

D $V = -6x^3 + 23x^2 + 16x - 48$

15. Kami was simplifying the expression $(2x - 3)(4x + 5)$ on a math exam. She got the answer $8x - 15$ but that answer was not one of the choices. What mistake did Kami make in simplifying and what should her answer have been?

F She should have added like terms to get $6x + 2$.

G She didn't fully distribute to get $8x^2 - 2x - 15$.

H She didn't combine middle terms correctly to get $8x^2 + 22x - 15$.

J She made a sign error and should have gotten $8x + 15$.

DIVIDING POLYNOMIALS

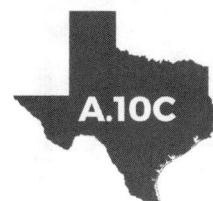

A.10C

The student is expected to determine the quotient of a polynomial of degree one and polynomial of degree two when divided by a polynomial of degree one and polynomial of degree two when the degree of the divisor does not exceed the degree of the dividend.

ⓘ TELL ME MORE...

You can divide polynomials using methods similar to how you would divide decimal numbers. Two commonly used methods to divide polynomials are factoring and the standard algorithm for division.

$$\text{Dividend} \div \text{Divisor} = \text{Quotient}$$

DIVISION BY FACTORING

Division by factoring is similar to reducing a fraction. Write the quotient as a fraction with the dividend as the numerator and the divisor as the denominator. Then, completely factor both polynomials. Common factors between the numerator and denominator divide to 1.

$$(2x^2 + 5x - 3) \div (2x - 1) = \frac{2x^2 + 5x - 3}{2x - 1}$$

$$= \frac{(2x - 1)(x + 3)}{2x - 1}$$

$$= \frac{2x - 1}{2x - 1} \cdot \frac{x + 3}{1}$$

$$= 1 \cdot \frac{x + 3}{1}$$

$$= x + 3$$

STANDARD ALGORITHM

The standard algorithm that you use for dividing whole numbers or decimals can be extended for use with polynomials. Represent the division with a division bracket where the dividend is placed inside the bracket and the divisor is placed in front of the bracket. The quotient, or answer, will be placed above the bracket in the appropriate place value.

> How many times will $2x$ divide into $2x^2$?

> Multiply x by $2x$ and x by -1. Align the x^2 and x terms. Subtract.

> Repeat the process.

> Bring down the next term.

$$\begin{array}{r} x + 3 \\ 2x - 1 \overline{) 2x^2 + 5x - 3} \\ \underline{-(2x^2 - x)} \\ 6x - 3 \\ \underline{-(6x - 3)} \\ 0 \end{array}$$

✓ EXAMPLES

EXAMPLE 1: Divide by factoring: $(5x^2 - 27x + 28) \div (x - 4)$.

STEP 1 Write the division sentence as a fraction. Place the dividend in the numerator and the divisor in the denominator.

$$\frac{5x^2 - 27x + 28}{x - 4}$$

STEP 2 Factor completely the numerator and denominator.

$$\frac{5x^2 - 27x + 28}{x - 4} = \frac{(5x - 7)(x - 4)}{x - 4}$$

$$\frac{(5x - 7)(x - 4)}{x - 4}$$

STEP 3 Look for common factors in the numerator and denominator. A common factor in the numerator and denominator divides to 1.

$$\frac{(5x-7)(x-4)}{x-4}$$

> $x-4$ is a common factor.

$$\frac{(5x-7)}{1} \cdot \frac{(x-4)}{x-4}$$

$$(5x-7) \cdot 1$$

5x – 7

EXAMPLE 2: Divide using the standard algorithm: $(3x^2 - 10x - 20) \div (x - 5)$.

STEP 1 Represent the division with a division bracket. Write the dividend inside the bracket and the divisor in front of the bracket.

$$x-5\overline{)3x^2-10x-20}$$

STEP 2 Divide $3x^2$ by x. Record the quotient above the $3x^2$ term.

$$\begin{array}{r} 3x \\ x-5\overline{)3x^2-10x-20} \end{array}$$

STEP 3 Multiply $3x$ by both terms in the divisor. Record the products beneath the appropriate term in the dividend.

$$\begin{array}{r} 3x \\ x-5\overline{)3x^2-10x-20} \\ 3x^2-15x \end{array}$$

STEP 4 Subtract the products. Bring down the next term.

$$\begin{array}{r} 3x \\ x-5\overline{)3x^2-10x-20} \\ -(3x^2-15x) \\ \hline 5x-20 \end{array}$$

STEP 5 Repeat the process.

$$\begin{array}{r} 3x\quad +5 \\ x-5\overline{)3x^2-10x-20} \\ -(3x^2-15x) \\ \hline 5x-20 \\ -(5x-25) \\ \hline 5 \end{array}$$

STEP 6 Record the remainder as a fraction using the remainder as the numerator and divisor as the denominator.

$$3x+5+\frac{5}{x-5}$$

EXAMPLE 3: Divide $12x - 15$ by $3x$.

STEP 1 To divide by factoring, write the division as a fraction with the dividend in the numerator and the divisor in the denominator.

$$\frac{12x - 15}{3x}$$

STEP 2 Factor completely the numerator and denominator.

$$\frac{12x - 15}{3x} = \frac{3(4x - 5)}{3x}$$

$$\frac{3(4x - 5)}{3x}$$

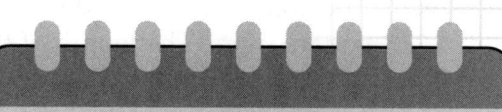

MAKE A NOTE . . .

If you had divided $12x - 15$ by $3x$ using the standard algorithm, would the quotient have been different? Explain.

STEP 3 Look for common factors in the numerator and denominator. A common factor in the numerator and denominator divides to 1.

3 is a common factor.

$$\frac{3(4x - 5)}{3x}$$

$$\frac{3}{3} \cdot \frac{(4x - 5)}{x}$$

$$1 \cdot \frac{(4x - 5)}{x}$$

$$\frac{4x - 5}{x}$$

STEP 4 Simplify the expression, if possible.

$$\frac{4x - 5}{x} = \frac{4x}{x} - \frac{5}{x} = 4 \cdot \frac{x}{x} - \frac{5}{x} = 4 \cdot 1 - \frac{5}{x}$$

$$4 - \frac{5}{x}$$

PRACTICE

Perform the division indicated using either factoring or the standard algorithm.

1. $(3x^2 - 8x + 5) \div (x - 1)$

2. $\dfrac{2x^2 + 9x - 18}{x + 6}$

3. $(12x^2 - 20x) \div 4x$

4. $\dfrac{x^2 - 36}{x^2 - 12x + 36}$

5. The area of a rectangle is $(12 - 9x)$ square units. The length of the rectangle is 3 units. What expression represents the width of the rectangle?

6. The combined rectangular area shown in the diagram is $6x^2 + 7x + 2$.

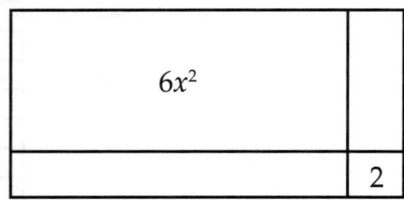

One of the side lengths of the rectangle is represented by the expression $(3x + 2)$. What expression represents the other dimension of the rectangle?

7. Caroline was dividing the polynomial $3x^2 - 27$ by the factor $(x - 3)$. Her work is shown below.

$$
\begin{array}{r}
3x + 9 \\
x - 3{\overline{\smash{\big)}\,3x^2 - 27}} \\
\underline{-(3x^2)} \\
0 - 27 \\
\underline{-(-27)} \\
0
\end{array}
$$

Evaluate Caroline's work and determine what, if any, mistake was made.

8. What is the remainder when $x^2 - 5x + 7$ is divided by $x + 1$?

9. What is the quotient when $2x^2 + 6x - 20$ is divided by $2x - 4$?

A $x + 5$

B $2x^2 + 4x - 16$

C $x - 1$ remainder -16

D $x - 10$

10. When the expression $5x^2 - 4x + 10$ is divided by the factor $(x + 1)$, what is the remainder?

F $5x - 9$

G   19

H $\dfrac{19}{x + 1}$

J   there is no remainder

APPLYING THE DISTRIBUTIVE PROPERTY

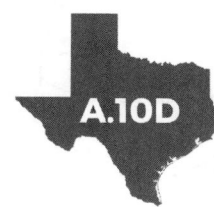

A.10D The student is expected to to rewrite polynomial expressions of degree one and degree two in equivalent forms using the distributive property.

ⓘ TELL ME MORE...

The **distributive property** of multiplication over addition states that if you have a sum of several quantities that is being multiplied by a number or expression, then you get the same result if you multiply that number or expression by each quantity and then add the products. In other words,

$$a(b + c + d) = ab + ac + ad$$

The distributive property also works for multiplication over subtraction.

$$a(b - c - d) = ab - ac - ad$$

> **Distributive property**
>
> Let a, b, and c represent real numbers.
>
> $$a(b + c) = ab + ac$$
>
> $$a(b - c) = ab - ac$$

You can also use the distributive property in reverse to factor an expression. If you have a sum or difference of several terms that all have a common factor, you can write that expression as a factor times a sum or difference. For example, the terms in the expression $20m^2n + 40m + 100n$ all have a common factor of 20.

$$20m^2n + 40m + 100n = 20(m^2n) + 20(2m) + 20(5n) = 20(m^2n + 2m + 5n)$$

✓ EXAMPLES

EXAMPLE 1: Simplify the expression $\frac{4}{9}(27m^2 - 18m + 90) + 4m^2 - 17$.

STEP 1 Apply the distributive property to multiply $\frac{4}{9}$ by each term inside the parentheses.

$$\frac{4}{9}(27m^2 - 18m + 90) + 4m^2 - 17$$

$$\frac{4}{9}(27m^2) + \frac{4}{9}(-18m) + \frac{4}{9}(90)$$

$$12m^2 - 8m + 40 + 4m^2 - 17$$

$12m^2 - 8m + 40 + 4m^2 - 17$

STEP 2 Combine like terms.

$$12m^2 - 8m + 40 + 4m^2 - 17$$

$$(12m^2 + 4m^2) - 8m + (40 - 17)$$

$$16m^2 - 8m + 23$$

$16m^2 - 8m + 23$

EXAMPLE 2: The surface area of a sphere can be calculated using the formula $A = 4\pi r^2$. The radius of a sphere is represented by the expression $\left(\frac{3}{8}x - 5\right)$ centimeters. Write an expression for the surface area of the sphere.

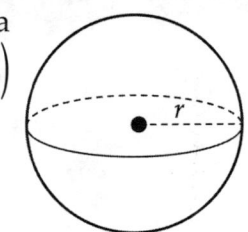

STEP 1 Substitute the expression $r = \frac{3}{8}x - 5$ into the surface area formula.

$$A = 4\pi r^2$$
$$A = 4\pi\left(\frac{3}{8}x - 5\right)^2$$

$$\mathbf{A = 4\pi\left(\frac{3}{8}x - 5\right)^2}$$

STEP 2 Follow the order of operations to simplify the expression. Begin with the exponent. Square the binomial.

$$A = 4\pi\left(\frac{3}{8}x - 5\right)^2$$
$$A = 4\pi\left(\frac{3}{8}x - 5\right)\left(\frac{3}{8}x - 5\right)$$
$$A = 4\pi\left(\frac{9}{64}x^2 - \frac{15}{4}x + 25\right)$$

$$\mathbf{A = 4\pi\left(\frac{9}{64}x^2 - \frac{15}{4}x + 25\right)}$$

STEP 3 Apply the distributive property. Multiply all three terms inside parentheses by 4π.

$$A = 4\pi\left(\frac{9}{64}x^2 - \frac{15}{4}x + 25\right)$$
$$A = 4\pi\left(\frac{9}{64}x^2\right) - 4\pi\left(\frac{15}{4}x\right) + 4\pi(25)$$
$$A = \frac{9\pi}{16}x^2 - 15\pi x + 100\pi$$

$$\mathbf{A = \frac{9\pi}{16}x^2 - 15\pi x + 100\pi}$$

YOU TRY IT!

Simplify: $4b\left(\frac{5}{4}b - 5\right) + 6\left(\frac{2}{3}b + 2\right)$.

Distribute $4b$ to both terms inside the adjacent parentheses and 6 to both terms the adjacent parentheses.

Combine like terms.

Expression: _____

EXAMPLE 3: Factor $36p^2q^2 - 72pq + 60p$ using the inverse of the distributive property.

STEP 1 Completely factor each of the three terms.
- $36p^2q^2 = 2 \cdot 2 \cdot 3 \cdot 3 \cdot p \cdot p \cdot q \cdot q$
- $-72pq = -1 \cdot 2 \cdot 2 \cdot 2 \cdot 3 \cdot 3 \cdot p$
- $60p = 2 \cdot 2 \cdot 3 \cdot 5 \cdot p$

STEP 2 Identify the common factors that appear in all three terms.
- $36p^2q^2 = \boxed{2} \cdot \boxed{2} \cdot \boxed{3} \cdot 3 \cdot \boxed{p} \cdot p \cdot q \cdot q$
- $-72pq = -1 \cdot \boxed{2} \cdot \boxed{2} \cdot 2 \cdot \boxed{3} \cdot 3 \cdot \boxed{p}$
- $60p = \boxed{2} \cdot \boxed{2} \cdot \boxed{3} \cdot 5 \cdot \boxed{p}$

2, 2, 3, and *p* are common to every term.

STEP 3   Divide each term by the greatest common factor, which is the product of all of the common factors.

$$\text{Greatest Common Factor} = 2 \cdot 2 \cdot 3 \cdot p = 12p$$

$$12p\left(\frac{36p^2q^2}{12p} - \frac{72pq}{12p} + \frac{60p}{12p}\right)$$

$$12p\left(\frac{12p}{12p} \cdot \frac{3pq^2}{1} - \frac{12p}{12p} \cdot \frac{6q}{1} + \frac{12p}{12p} \cdot \frac{5}{1}\right)$$

$$12p(1 \cdot 3pq^2 - 1 \cdot 6q + 1 \cdot 5)$$

$$12p(3pq^2 - 6q + 5)$$

12p(3pq² – 6q + 5)

PRACTICE

Simplify the following expressions using the distributive property.

1. $\frac{1}{3}x(x-12) + \frac{5}{6}(2x^2 + 42x) - \frac{3}{2}(4x^2 - 18x + 8)$

2. $17p^2 - \frac{3}{5}(15p - 6) - 9p(p-2)$

3. $a(a+2b-3c) + 3c(a-2b+3c) - 2b(a-b-c)$

Factor the following expressions using the distributive property.

4. $36m^2 - 60m$

5. $20a^2b^2 + 30a^2b - 50ab$

6. $16xy^2 + 48x^2y^2 + 56x^2y$

7. The expression below can be used to determine the measure of the third side of an isosceles triangle given the perimeter and the measure of the two congruent sides.

$$8\left(\frac{1}{2}x^2 + 3x - 7\right) - 2(0.75x(2x - 28))$$

What is the measure of the third side?

8. Jeremy's solution to a problem involving polynomial expressions is shown below.

Step 1: $a(2a + 7b) - b(9 + 5a^2 - 4a) + 6b^2$

Step 2: $2a^2 + 7ab - b(9 + 5a^2 - 4a) + 6b^2$

Step 3: $2a^2 + 7ab - 9b + 5a^2b - 4ab + 6b^2$

Step 4: $7a^2b + 3ab - 9b + 6b^2$

What mistakes, if any, did Jeremy make and in which steps were the errors first made?

9. Cynthia simplified an expression on her recent math test. Her steps are shown below.

Step 1: $6n\left(\dfrac{2}{3}m + 4\right) - 7(n - 2) - m(2n + 5)$

Step 2: $4mn + 24n - 7(n - 2) - m(2n + 5)$

Step 3: $4mn + 24n - 7n - 14 - m(2n + 5)$

Step 4: $4mn + 24n - 7n - 14 - 2mn + 5m$

Step 5: $17n + 2mn + 5m - 14$

What type of mistake did Cynthia make?

A Combined like terms incorrectly in Step 5

B Multiplication error in Step 2

C A sign error in Step 3

D A distribution error in Step 4

10. Which expression is NOT equivalent to $-2x(x - 7) - 3x(x + 3) + 10$?

F $-5(x - 2)(x + 1)$

G $3(7x - x^2) - 2(x^2 + 8x - 5)$

H $5(-x^2 + x - 2)$

J $\dfrac{2}{3}(3x^2 + 9x + 21) - \dfrac{1}{5}(20 + 5x + 35x^2)$

11. Which expression is equivalent to $18j^2k^2 - 27jk - 9k^2$?

A $9k(2j^2k - 3j - k)$

B $9j^2k(2k - 3 - k)$

C $9jk(2jk - 3 - k)$

D $9(2j^2k^2 - 3jk - 1)$

12. A student solved the following problem. In which step of the solution process does the student first make an error in simplifying the expression.

Step 1: $(3x - 4)(-2x - 5) + 7x(x - 3) + 8$

Step 2: $-6x^2 - 15x + 8x + 20 + 7x(x - 3) + 8$

Step 3: $-6x^2 - 15x + 8x + 20 + 7x^2 - 21x + 8$

Step 4: $13x^2 - 28x + 28$

A Step 1

B Step 2

C Step 3

D Step 4

FACTORING POLYNOMIALS

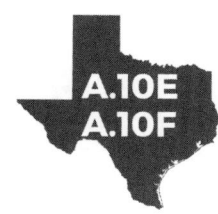

A.10E
A.10F

The student is expected to factor, if possible, trinomials with real factors in the form $ax^2 + bx + c$, including perfect square trinomials of degree two.

The student is expected to decide if a binomial can be written as the difference of two squares and, if possible, use the structure of a difference of two squares to rewrite the binomial

ⓘ TELL ME MORE...

Binomials are polynomials with two terms. **Trinomials** are polynomials with three terms. Sometimes, a polynomial expression can be written in terms of two or more factors that can then be multiplied together to generate the original expression. Doing so is called **factoring** the expression. Factoring an expression allows you to break the expression down and look for patterns.

Certain trinomials and binomials have characteristics that allow you to factor them more easily. For example, trinomials and binomials containing terms that are **perfect squares** have special patterns.

$$x^2 - 10x + 25 = (x - 5)(x - 5) = (x - 5)^2$$

$$9m^2 + 24m + 64 = (3m + 8)(3m + 8) = (3m + 8)^2$$

$$x^2 - 49 = (x + 7)(x - 7)$$

> **Perfect Square Trinomials**
> $a^2 + 2ab + b^2 = (a + b)^2$
> $a^2 - 2ab + b^2 = (a - b)^2$
>
> **Perfect Square Binomials**
> $a^2 - b^2 = (a + b)(a - b)$

The first pattern to look for when factoring a polynomial is a common factor among all of the terms.

$$4 \cdot 4 \cdot x$$
$$4x^2 - 16x + 40 = 4(x^2 - 4x + 10)$$
$$4 \cdot x \cdot x \qquad 4 \cdot 10$$

You can also factor trinomials using patterns in the whole number factors of the coefficients of the terms. This approach is commonly called **trial and error** or **guess and check** because you make an educated guess about which factor pairs to use and then check the multiplication to see if it worked.

Factor pair: 1 and 1 (implied 1)

$$a^2 + 12a + 20 = (a + 2)(a + 10)$$

$2(10) = 20$, the last term
$2 + 10 = 12$, the middle term

Factor pairs: 1 and 20, 2 and 10, 4 and 5

You can also look for groups of factors in a trinomial. This factoring technique is called **grouping** and is an inverse of the double-distribution procedure commonly used to multiply binomials.

$$2x^2 - 11x + 15$$

Break the middle term into two parts. $2 \times 15 = 30$, so use two factors of 30.

$$2x^2 - 6x - 5x + 15$$

$$2x(x - 3) - 5(x - 3)$$

Factor out a common factor from the first two terms and the last two terms.

Factor out the common binomial factor.

$$(2x - 5)(x - 3)$$

EXAMPLE 1: Completely factor $5m^2 - 35m - 300$.

STEP 1 Look for a common factor among all terms of the polynomial.

$$5m^2 - 35m - 300$$

$5 \cdot m \cdot m$	$5 \cdot 7 \cdot m$	$5 \cdot 60$

$$5(m^2) - 5(7m) - 5(60)$$

$5(m^2 - 7m - 60)$

STEP 2 Look to see if there are special patterns, such as perfect squares. m^2 is a perfect square ($m \cdot m$) but 60 is not, so this trinomial does not have a perfect square pattern. Instead, look for factor pairs.

- Multiply the coefficient of the m^2 term by the last term: $1 \cdot 60 = 60$
- $60 = 1 \cdot 60, 2 \cdot 30, 3 \cdot 20, 4 \cdot 15, 5 \cdot 12,$ and $6 \cdot 10$
- Since the last term is actually −60, look for a factor pair that subtracts to give −7, the coefficient of the middle term.
- Use 5 and −12: $5(-12) = -60$ and $5 - 12 = -7$.

$5(m + 5)(m - 12)$

STEP 3 Use the table feature of graphing technology to check the factors alongside the original polynomial. If the y-values for both expressions are all the same, then the expressions are equivalent.

$5m^2 - 35m - 300 = 5(m + 5)(m - 12)$

Plot1	Plot2	Plot3
∎\Y₁⊟5X²−35X−300		
∎\Y₂⊟5(X+5)(X−12)		
∎\Y₃=		
∎\Y₄=		
∎\Y₅=		
∎\Y₆=		
∎\Y₇=		
∎\Y₈=		

X	Y1	Y2
0	-300	-300
1	-330	-330
2	-350	-350
3	-360	-360
4	-360	-360
5	-350	-350
6	-330	-330
7	-300	-300
8	-260	-260
9	-210	-210
10	-150	-150

EXAMPLE 2: Completely factor $6x^2 - 29x + 35$.

STEP 1 Look for a common factor among all terms of the polynomial.

6, 29, and 35 have no common factors.

$6x^2 - 29x + 35$

STEP 2 Look to see if there are special patterns, such as perfect squares. $6x^2$ and 35 are not perfect squares. Instead, use a grouping strategy.

- Multiply the coefficient of the x^2 term by the last term: $6 \cdot 35 = 210$
- $210 = 1 \cdot 210, 2 \cdot 105, 3 \cdot 70, 5 \cdot 42, 6 \cdot 35, 7 \cdot 30, 10 \cdot 21,$ and $14 \cdot 15$
- Since the middle term is −29, look for a factor pair that sums to −29. Both factors will then be negative.
- Use −14 and −15: $-14 + (-15) = -29$
- Split the middle term, $-29x$, into $-14x$ and $-15x$.

$6x^2 - 14x - 15x + 35$

STEP 3 Factor out a common factor in the first two terms. Factor out a common factor in the last two terms.

$$(6x^2 - 14x) - (15x - 35)$$
$$2x(3x - 7) - 5(3x - 7)$$

$2x(3x - 7) - 5(3x - 7)$

STEP 4 Factor out the common binomial.

$$2x(3x - 7) - 5(3x - 7)$$
$$(2x - 5)(3x - 7)$$

$(2x - 5)(3x - 7)$

EXAMPLE 3: Completely factor $z^2 - 121$.

STEP 1 Look for a common factor among all terms of the polynomial. 1 and 121 have no common factors.

$z^2 - 121$

STEP 2 Look to see if there are special patterns, such as perfect squares. z^2 ($z \cdot z$) and 121 ($11 \cdot 11$) are both perfect squares. Use the binomial square formula.

$$a^2 - b^2 = (a + b)(a - b)$$
$$z^2 - 121 = z^2 - 11^2$$
$$(z + 11)(z - 11)$$

$(z + 11)(z - 11)$

EXAMPLE 4: Completely factor $25d^2 - 70d + 49$.

STEP 1 Look for a common factor among all terms of the polynomial.

25, 70, and 49 have no common factors.

$25d^2 - 70d + 49$

STEP 2 Look to see if there are special patterns, such as perfect squares. $25d^2$ ($5d \cdot 5d$) and 49 ($7 \cdot 7$) are both perfect squares. Use the trinomial square formula with a negative middle term.

$$a^2 - 2ab + b^2 = (a - b)^2$$
$$25d^2 - 70d + 49 = (5d)^2 - 2(5d)(7) + 7^2$$
$$(5d - 7)^2$$

$(5d - 7)^2$

YOU TRY IT! #1

Fill in the blanks to factor $3n^2 - 13n - 10$.

$$3n^2 - 13n - 10$$

$$3n^2 - 15n + \underline{\hspace{1cm}} - 10.$$

$$3n(\underline{\hspace{0.7cm}} - \underline{\hspace{0.7cm}}) + 2(n - \underline{\hspace{0.7cm}})$$

$$(3n + 2)(\underline{\hspace{0.7cm}} - \underline{\hspace{0.7cm}})$$

Factored form: _____

YOU TRY IT! #2

Completely factor $2x^2 + 12x - 32$.

Common factor? _____

Perfect square pattern? _____

Factored pairs: _____

_____

Factored form: _____

Factor completely.

1. $16x^2 - 40x + 25$

3. $8 + 23x - 3x^2$

5. $64 - 9m^2$

2. $3x^2 - 45x + 150$

4. $a^2 - 256$

6. $z^2 + 22z + 121$

Solve each problem below.

7. When the expression $3x^2 - 5x - 12$ is written as the product of two factors, $(x - 3)$ is one of the factors What is the other factor?

9. The area of a rectangle is $196x^2 - 625$. What are the dimensions of the rectangle?

8. A polynomial $9x^2 + \underline{\ ?\ } + 64$ is a perfect square polynomial. What is the missing middle term that allows the polynomial to be written as a squared factor?

10. Devonte and Marquis are factoring polynomial expressions. When they get to the expression $5y^2 - 20$, Devonte says the problem is factorable as a difference of two squares. Marquis disagrees and says the expression is prime and cannot be factored. Which student is correct and why?

11. Nguyen and Patrice are solving factoring problems involving difference of squares. They have been assigned to demonstrate the solution to $27n^2 - 12$. Patrice says the problem is not a difference of squares problem and is therefore prime. How can Nguyen demonstrate the problem is a difference of squares problem? What is the equivalent expression in factored form?

12. The area of a rectangular swimming pool is $2x^2 + x - 45$ square meters as shown in the diagram.

What are the dimensions of the swimming pool?

13. Which expression is a factor of $6x^2 + 21x + 15$?

A $(6x + 5)$

B $(3x + 5)$

C $(2x + 5)$

D $(2x + 3)$

14. Which of the following polynomial expressions can be factored as the difference of two squares?

F $x^2 - 8x + 16$

G $4x^2 + 25$

H $36 - 81x^2$

J $49x - 16$

15. Which expression is equivalent to $8n^2 - 18n + 7$?

A $(4n - 7)(2n - 1)$

B $(4n + 7)(2n + 1)$

C $(2n - 7)(4n - 1)$

D $(2n + 7)(4n + 1)$

SIMPLIFYING RADICAL EXPRESSIONS

A.11A The student is expected to simplify numerical radical expressions involving square roots.

(i) TELL ME MORE...

A **radical** expression is an expression involving a square root symbol. The square root is the inverse operation of squaring a number, so square roots are related to numbers and variables that are squared.

Squaring a number means to multiply the number by itself. Taking the square root of a number means to look for a number that, when multiplied by itself, gives you the original number. A number, such as 9, 16, 25, or 36, that has a whole number square root is called a **perfect square**.

> **Square Roots**
> $\sqrt{a^2} = a$

$$3 \times 3 = 9, \text{ so } \sqrt{9} = 3 \quad 4 \times 4 = 16, \text{ so } \sqrt{16} = 4 \quad 5 \times 5 = 25, \text{ so } \sqrt{25} = 5 \quad 6 \times 6 = 36, \text{ so } \sqrt{36} = 6$$

Not all numbers are perfect squares. You can still simplify a radical expression involving a number that isn't a perfect square if you look for factor pairs within that number.

For example, 162 is not a perfect square. But its prime factorization contains factor pairs.

$$162 = 2 \times 3 \times 3 \times 3 \times 3$$
$$162 = 2 \times 9 \times 9$$

Since 162 contains pair of 9's as factors, you can use that to simplify $\sqrt{162}$.

$$\sqrt{162} = \sqrt{2 \times 9 \times 9} = \sqrt{2} \times \sqrt{9 \times 9} = \sqrt{2} \times \sqrt{9^2} = \sqrt{2} \times 9 = 9\sqrt{2}$$

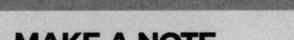

> **MAKE A NOTE . . .**
>
> List several perfect squares:
>
> 1, 4, 9, . . .

✓ EXAMPLES

EXAMPLE 1: Simplify $\sqrt{384}$.

STEP 1 List the factors of 384. Look for perfect squares that are factors of 384.

384

1	384
2	192
3	128
4	96
6	64
8	48
12	32
16	24

STEP 2 Identify the greatest perfect square that is a factor of 384.

- 64 is the greatest perfect square factor of 384.
- $64 \times 6 = 384$

64

STEP 3 Write $\sqrt{384}$ as the square root of 64×6.

$\sqrt{64 \times 6}$

384	
1	384
2	192
3	128
4	96
6	64
8	48
12	32
16	24

STEP 4 Bring the perfect square outside the radical. When you do, take its square root.

$$\sqrt{64} \times 6$$
$$\sqrt{64} \times \sqrt{6}$$
$$8\sqrt{6}$$

8√6

EXAMPLE 2: Simplify $\sqrt{96}$.

STEP 1 Use a factor tree to determine the prime factorization of 96.

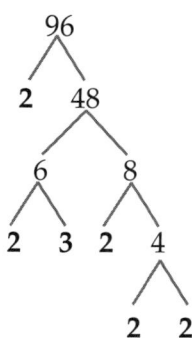

$96 = 2 \times 2 \times 2 \times 2 \times 2 \times 3 = 2^5 \times 3$

> ## YOU TRY IT!
>
> Simplify $\sqrt{375}$ completely.
>
> Factors of 375:
>
>
> Greatest perfect square factor:_____
>
> $\sqrt{375} = \sqrt{\square \times \square} = \square\sqrt{\square}$
>
> Simplified:_____

STEP 2 Write $\sqrt{96}$ as the square root of its prime factorization. Circle pairs of prime factors.

$$\sqrt{96} = \sqrt{(2 \times 2) \times (2 \times 2) \times 2 \times 3}$$

STEP 3 For each pair of prime factors, write one of the factors (which is the square root of the product of the two factors) in front of the radical. Record remaining prime factors beneath the radical.

$$\sqrt{96} = \sqrt{(2 \times 2) \times (2 \times 2) \times 2 \times 3}$$
$$\sqrt{96} = 2 \times 2\sqrt{2 \times 3}$$
$$\sqrt{96} = 4\sqrt{6}$$

4√6

EXAMPLE 3: The area of a square is 504 square centimeters. What is the side length of the square, in simplified radical form? (Hint: for a square, $A = s^2$.)

STEP 1 The side length of a square is the square root of its area. Write an expression showing the side length as the square root of the area, 504 square centimeters.

$$s = \sqrt{504}$$

$A = s^2$

$\sqrt{A} = \sqrt{s^2}$

$\sqrt{A} = s$

STEP 2 Write the prime factorization of 504.

$$504 = 2 \times 2 \times 2 \times 3 \times 3 \times 7$$

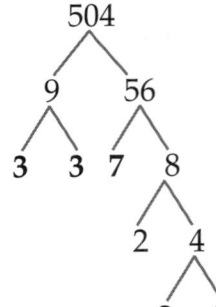

STEP 3 Write $\sqrt{504}$ as the square root of its prime factorization. Circle pairs of prime factors.

$$\sqrt{504} = \sqrt{(2 \times 2) \times 2 \times (3 \times 3) \times 7}$$

STEP 4 For each pair of prime factors, write one of the factors (which is the square root of the product of the two factors) in front of the radical. Record remaining prime factors beneath the radical.

$$\sqrt{504} = \sqrt{(2 \times 2) \times 2 \times (3 \times 3) \times 7}$$

$$\sqrt{504} = 2 \times 3\sqrt{2 \times 7}$$

$$\sqrt{504} = 6\sqrt{14}$$

The side length of the square is $6\sqrt{14}$ centimeters.

PRACTICE

Completely simplify.

1. $\sqrt{288}$

2. $\sqrt{\dfrac{200}{49}}$

3. $4\sqrt{168}$

4. Johan was using the discriminant of the quadratic formula to determine the number of real roots for the quadratic equation he was solving. The discriminant for his equation was $\sqrt{7^2 - 4(2)(3)}$. Write an expression, in simplest form, that is equivalent to Johan's discriminant.

5. The area of a square garden is 175 square yards. What is the measure of the side length of the square garden written in simplest radical form?

6. Palomo is using square paving stones to make a walkway. Each stone has a side length of $4\sqrt{3}$ inches. What is the area of each stone written in simplest form?

7. The area of a square sandbox on a playground is 864 ft². The city plans to put a new barrier around the outer edge of the sandbox. How many feet of barrier, written in simplest form, will they need for one side of the sandbox?

8. Mika was simplifying the radical expression $\sqrt{675}$. Her work is shown.

Step 1: $\sqrt{675}$

Step 2: $\sqrt{25 \times 27}$

Step 3: $\sqrt{(5 \times 5) \times (9 \times 3)}$

Step 4: $\sqrt{(5 \times 5) \times (3 \times 3 \times 3)}$

Step 5: $8\sqrt{3}$

What mistake, if any, did Mika make in simplifying the radical expression and in which step did she make the error?

9. The volume, V, in cubic units of a cylinder is given by the formula $V = \pi r^2 h$, where h is the height and r is the radius of the cylinder. What is the radius of a cylinder with a height of 6 cm and a volume of $1{,}152\pi$ cm³?

A 192 cm

B $8\sqrt{3}$ cm

C $12\sqrt{3}$

D $64\sqrt{3}$

10. Which expression is equivalent to $\sqrt{980}$?

F $5\sqrt{14}$

G $196\sqrt{5}$

H $14\sqrt{5}$

J $20\sqrt{7}$

LAWS OF EXPONENTS

A.11B

The student is expected to simplify numeric and algebraic expressions using the laws of exponents, including integral and rational exponents.

ⓘ TELL ME MORE...

In addition to properties of algebra, you can use the **laws of exponents** to simplify numerical and algebraic expressions.

Some rules, such as multiplication and division, require the bases of two different terms to be alike. Other rules, such as power to a power, negative exponent, zero exponent, and rational exponent, require one term.

Multiplication with Like Bases

$$a^m a^n = a^{m+n}$$

Division with Like Bases

$$\frac{a^m}{a^n} = a^{m-n}$$

Power to a Power

$$(a^m)^n = a^{mn}$$

Negative Exponent

$$a^{-m} = \frac{1}{a^m}$$

Zero Exponent

$$a^0 = 1$$

Rational Exponent

$$\sqrt[n]{a^m} = a^{\left(\frac{m}{n}\right)}$$

✓ EXAMPLES

EXAMPLE 1: The length of a rectangle is $7.5a^4b^{\left(\frac{5}{4}\right)}c^{-4}$ meters and the width of a rectangle is $2.5a^8b^{\left(\frac{3}{4}\right)}c^5$ meters. Write an expression for the area of the rectangle in square meters. (Hint: For a rectangle, $A = lw$.)

STEP 1 The area of a rectangle is the product of its length and width. Multiply the two expressions to determine the area.

$$\text{Area} = \text{length} \times \text{width}$$
$$\text{Area} = 7.5a^4b^{\left(\frac{5}{4}\right)}c^{-4} \times 2.5a^8b^{\left(\frac{3}{4}\right)}c^5$$

$$\mathbf{7.5a^4b^{\left(\frac{5}{4}\right)}c^{-4} \times 2.5a^8b^{\left(\frac{3}{4}\right)}c^5}$$

STEP 2 Multiply the coefficients like usual. Use the multiplication property of exponents to add the powers for each base, a, b, and c.

$$7.5a^4b^{\left(\frac{5}{4}\right)}c^{-4} \times 2.5a^8b^{\left(\frac{3}{4}\right)}c^5$$
$$18.75a^{(4+8)}b^{\left(\frac{5}{4}+\frac{3}{4}\right)}c^{(-4+5)}$$
$$18.75a^{(12)}b^{(2)}c^{(1)}$$

$$\mathbf{18.75a^{12}b^2c}$$

EXAMPLE 2: The density of a substance is its mass divided by its volume $(d = \frac{m}{V})$. The mass of a particular mineral is $4(x^2)^3x^5$ grams and the volume is $8(x^6)^3$ cubic centimeters, write an expression for the density of the mineral.

STEP 1 Write the quotient for density using the expressions for mass and volume.

$$d = \frac{m}{V}$$
$$d = \frac{4(x^2)^3x^5}{8(x^6)^3}$$

$$\mathbf{d = \frac{4(x^2)^3x^5}{8(x^6)^3}}$$

STEP 2 Simplify the numerator using the power-to-a-power property and the multiplication property.

$$d = \frac{4(x^2)^3x^5}{8(x^6)^3}$$
$$d = \frac{4x^6x^5}{8(x^6)^3}$$
$$d = \frac{4x^{11}}{8(x^6)^3}$$

$$\mathbf{d = \frac{4x^{11}}{8(x^6)^3}}$$

STEP 3 Simplify the denominator using the power-to-a-power property.

$$d = \frac{4x^{11}}{8(x^6)^3}$$
$$d = \frac{4x^{11}}{8x^{18}}$$

$$\mathbf{d = \frac{4x^{11}}{8x^{18}}}$$

STEP 4 Simplify the quotient by dividing the coefficients and using the division property of exponents.

$$d = \frac{4x^{11}}{8x^{18}}$$
$$d = \frac{1x^{11-18}}{2}$$
$$d = \frac{x^{-7}}{2}$$

$$\mathbf{d = \frac{x^{-7}}{2}}$$

STEP 5 Rewrite the negative exponent as the reciprocal of the base and power.

$$d = \frac{x^{-7}}{2} = \frac{1}{2x^7}$$

$$\mathbf{d = \frac{1}{2x^7}}$$

YOU TRY IT!

Simplify $4(x^2)^{\frac{3}{8}}(x^5)^{\frac{7}{10}}$.

Apply the power-to-a-power rule to each base.

Apply the multiplication rule.

Simplified expression: _____

EXAMPLE 3 The area of a circle is found using the formula $A = \pi r^2$, where r is the radius of the circle. The radius of a circle is $3x^5y^8$ centimeters and the area is equivalent to $9\pi x^n y^p$ square centimeters. What is the value of n? Record your answer and fill in the bubbles on your answer document.

STEP 1   Write an expression for the circle's area using the expression for the radius.

$$A = \pi r^2 = \pi(3x^5y^8)^2$$

$$\boldsymbol{A = \pi(3x^5y^8)^2}$$

STEP 2   Apply the power-to-a-power property to simplify the expression for area.

$$A = \pi(3x^5y^8)^2$$

$$A = \pi(3^2)(x^5)^2(y^8)^2$$

$$A = 9\pi x^{5(2)}y^{8(2)}$$

$$A = 9\pi x^{10}y^{16}$$

$$\boldsymbol{A = 9\pi x^{10}y^{16}}$$

STEP 3   The expression for area is equivalent to $9\pi x^n y^p$. So, the exponent of x is equivalent to n and the exponent of y is equivalent to p. Identify the value of n. Practice using the grid with the instructions.

1. Record a 1 in the first column containing numbers. Record a 0 in the next column.

2. Bubble the 1 beneath the numeral 1. Bubble the 0 beneath the numeral 0.

PRACTICE

Simplify each of the following expressions so that there are only positive exponents.

1. $\dfrac{96m^6n^5p^2}{8m^3n^8p^3}$

3. $4(x^3)^{\frac{5}{3}}\left(3x^{\frac{1}{2}}\right)^4$

2. $\left(n^{-\frac{3}{2}} \times n^{\frac{1}{2}}\right)^4$

4. $16a^6b^{-4}c \times \dfrac{7}{8}a^{14}b^3c^8$

5. The dimensions of a rectangular window are $2x^2y^3$ and $5x^3y^5$. What is the area of the window? $[A = lw]$

6. The area of a triangular college pennant is $4x^6y^4$ square units. If the height of the pennant is $4x^4y^3$ units, what is the length of the base in units? $[A = \frac{1}{2}bh]$

7. The radius of a soda can is $8x^2y^4$ and the height of the can is $10x^3y^5$. What is the volume of the can in terms of pi? $[V = \pi r^2 h]$

8. A circular swimming pool has a radius of $3xy^2z^3$ and a depth of $4x4^{yz5}$. How much water would be needed to completely fill the pool? $[V = \pi r^2 h]$

9. A rectangular garden has a length of $2x^4y^2$ and a width of $4xy^3$. The garden will be planted with 8 different types of plants and each will have equal area in the garden. What amount of area will be devoted to each type of plant? $[A = lw]$

10. A terrarium has a base width of $2xy^2z^4$ and a length of $3x^3y^4z^{-2}$. The volume of the terrarium equals $30x^6y^{11}z^{18}$. What is the height of the terrarium? $[V = lwh]$

A $180x^{10}y^{17}z^{20}$

B $5x^3y^3z^{10}$

C $6x^2y^5z^{16}$

D $5x^2y^5z^{16}$

11. Which expression is equivalent to $\dfrac{(-8a^4b^5)(3a^3b^{-2})}{(-4a^5b^6c^3)^2}$?

F $\dfrac{-3}{2a^3b^9c^6}$

G $\dfrac{-3}{a^3b^9c^6}$

H $\dfrac{6a^2}{b^3c^3}$

J $\dfrac{6a^2}{b^6c^6}$

12. The expression $(b^6)(b^{-4})^2$ is equivalent to bn. What is the value of n? Record your answer and fill in the bubbles?

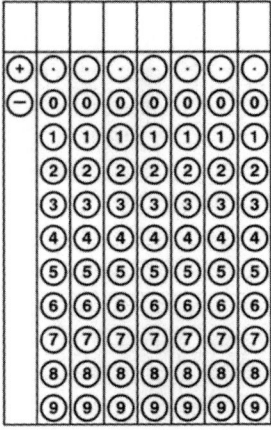

RELATIONS AND FUNCTIONS

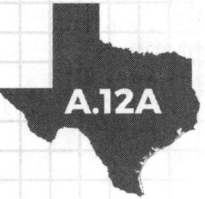

A.12A The student is expected to decide whether relations represented verbally, tabularly, graphically, and symbolically define a function.

ⓘ TELL ME MORE...

A **relation** is a particular relationship between two sets of numbers. A **function** is a relation in which each value of the independent variable (input value) generates only one value of the dependent variable (output value). In other words, each element of the domain of a function generates ("maps to") only one element of the range of the function. In the mapping shown, x represents the independent variable and y represents the dependent variable. Notice how each value of x maps to one value of y; y is a function of x and this relation is a function.

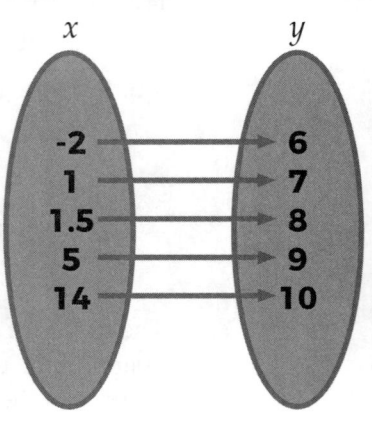

You can represent relations and functions with verbal descriptions, tables, graphs, and symbols. A functional relationship has only one output value for each input value. The examples below have x as the independent variable and y as the dependent variable.

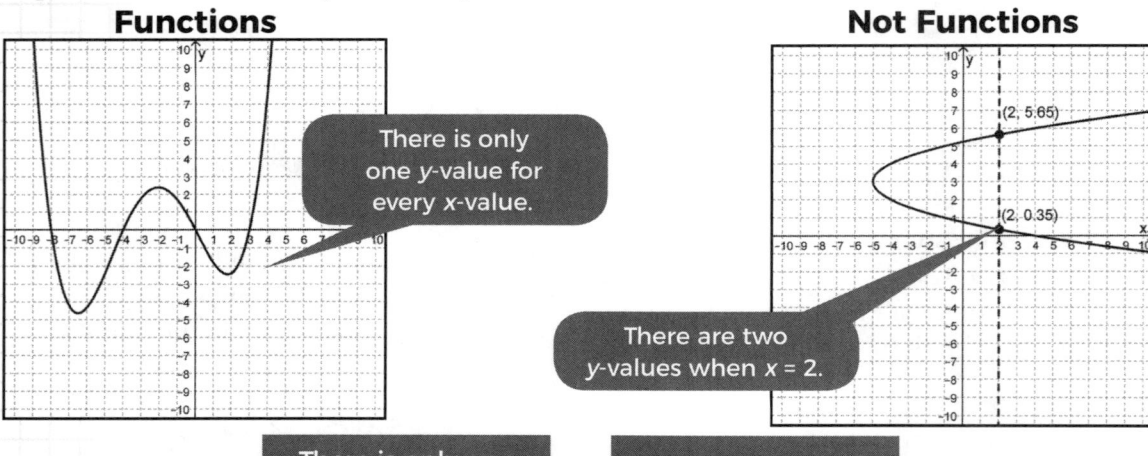

Functions

There is only one y-value for every x-value.

Not Functions

There are two y-values when $x = 2$.

There is only one y-value for every x-value.

$y = (x - 3)^2 + 5$

There are two y-values when $x = 2$.

$y = (x - 3)^2 - 5$

x	1	2	3	4
y	10	12	14	16

There is only one y-value for every x-value.

x	1	2	1	4
y	10	12	14	16

When $x = 1$, $y = 10$ or $y = 14$. There are two possible y-values.

Deirdre records the amount of rainfall, in inches, over a 3-hour period.

Duc records the amount of time it takes 20 classmates to get to school and the distance they live from school.

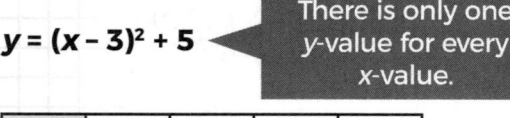

There is only one rainfall measurement for each moment in time.

Students could live different distances from school and still take the same time to get there.

EXAMPLES

EXAMPLE 1: Darnell and his science lab group collected two rounds of data where x represents the independent variable and y represents the dependent variable. Each round is shown in a table below. Which table represents y as a function of x? Which table does not represent y as a function of x? Explain how you know.

Round 1			Round 2	

x	y
3	10.3
4	15.6
5	19.7
4	15.6
6	19.7

x	y
4	15.6
4	15.9
5	19.5
6	21.3
7	26.8

STEP 1 Look at the table for Round 1 to see if any x-values are repeated. If so, identify the ordered pairs (x, y) representing the rows in the table.

Round 1 has x = 4 twice: (4, 15.6) in the second row and (4, 15.6) in the fourth row

Round 1

x	y
3	10.3
4	15.6
5	19.7
4	15.6
6	19.7

STEP 2 If the corresponding y-values are the same, the relation is a function. If not, the relation is not a function.

- For (4, 15.6) and (4, 15.6), y = 15.6 whenever x = 4.

Round 1 represents a function because each x-value generates only one y-value.

STEP 3 Look at the table for Round 2 to see if any x-values are repeated. If so, identify the ordered pairs (x, y) representing the rows in the table.

Round 2 has x = 4 twice: (4, 15.6) in the first row and (4, 15.9) in the second row.

Round 2

x	y
4	15.6
4	15.9
5	19.5
6	21.3
7	26.8

STEP 4 If the corresponding y-values are the same, the relation is a function. If not, the relation is not a function.

- For (4, 15.6) and (4, 15.9), y is not the same whenever x = 4.

Round 2 does not represent a function because one x-value generates more than one y-value.

EXAMPLE 2: Does the graph represent y as a function of x? How do you know?

STEP 1 Look at the graph to see if there are any x-values that have multiple y-values.

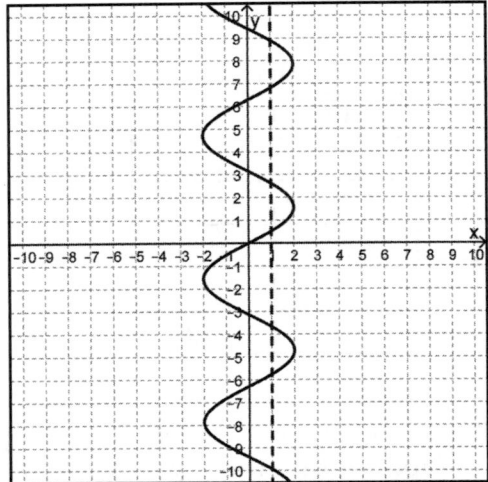

When x = 1, there are seven possible y-values.

STEP 2 If there are multiple y-values (dependent variable values) for any x-value (independent variable value), then the relation is not a function. Otherwise, the relation is a function.

The relation is not a function because there are multiple values of the dependent variable for one value of the independent variable.

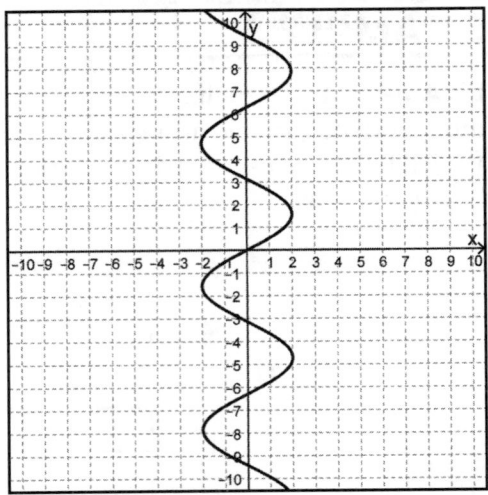

On a graph, the independent variable is on the horizontal (x) axis, so a vertical line represents all of the possible points with that x-value. If a vertical line crosses the graph more than once, then the relation is not a function. Hence, this procedure is called the vertical line test.

YOU TRY IT!

Does the graph below represent y as a function of x? How do you know?

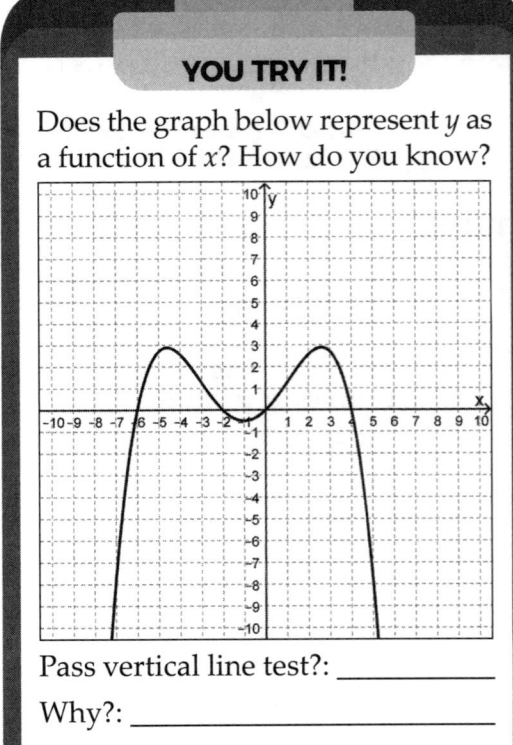

Pass vertical line test?: _____

Why?: _____

_____

EXAMPLE 3 Which of the following sets of points represents y as a function of x?

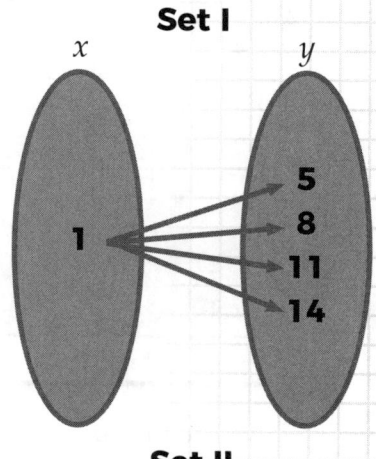

I. $\{(1, 5), (1, 8), (1, 11), (1, 14)\}$

II. $\{(2, 5), (3, 5), (4, 5), (5, 5)\}$

III. $\{(3, 10), (4, 15), (5, 20), (6, 25)\}$

STEP 1 Use a mapping to determine if Set I represents a function. If it does, then each x-value will generate the same y-value every time.

- In the mapping, x generates 4 different y-values.

Set I is not a function.

STEP 2 Use a mapping to determine if Set II represents a function. If it does, then each x-value will generate the same y-value every time.

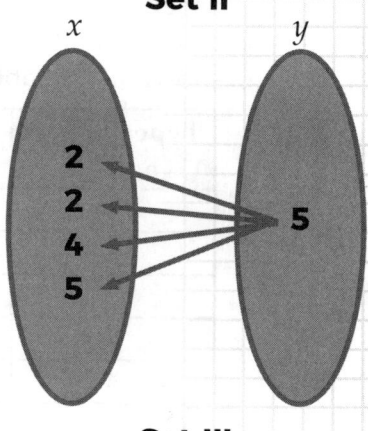

- In the mapping, each x generates only one y-value, even if it's the same y-value.

Set II is a function.

STEP 3 Use a mapping to determine if Set III represents a function. If it does, then each x-value will generate the same y-value every time.

- In the mapping, each x generates only one y-value, even if it's the same y-value.

Set III is a function.
Sets II and III represent y as a function of x.

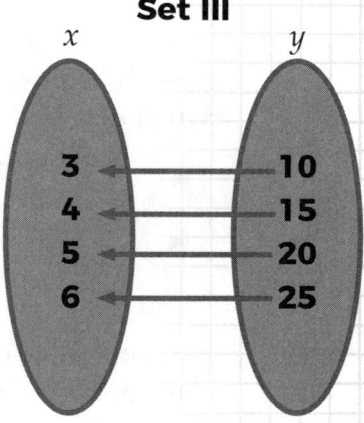

PRACTICE

For questions 1-6, determine if the graph or table represents y as a function of x.

1.

x	y
12	100
14	150
16	100
18	250
20	300

2.

x	y
15	40.7
16	71.5
17	85.4
16	71.5
17	89.3

3.

x	y
-10	71
-8	65
-10	71
-4	52
-2	38

4.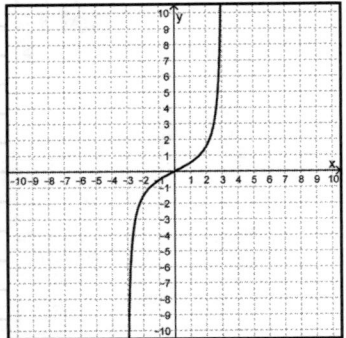

5.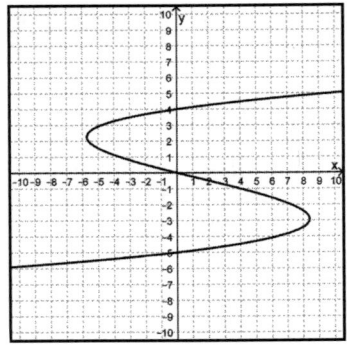

6.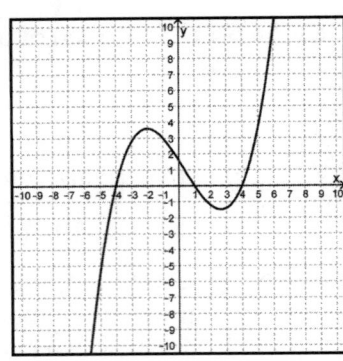

---

7. The number of fish caught by fishermen based on the number of different bait types used for fishing is shown in the table.

Types of bait used, x	Number of fish caught, y
1	7
3	10
7	9
7	7
4	5

Is the number of fish caught a function of the number of different types of bait used? Explain why or why not.

8. A plant fertilizer contains phosphorus-32, which has a half-life of 14.3 days. A plant is sprayed with a liquid version of the fertilizer that contains 84 milligrams of phosphorus-32. Why is the amount of phosphorus-32 available to the plant after being sprayed with the fertilizer a function of the number of intervals of 14.3 days?

9. Moana plotted the graph shown below and explained that it is a function.

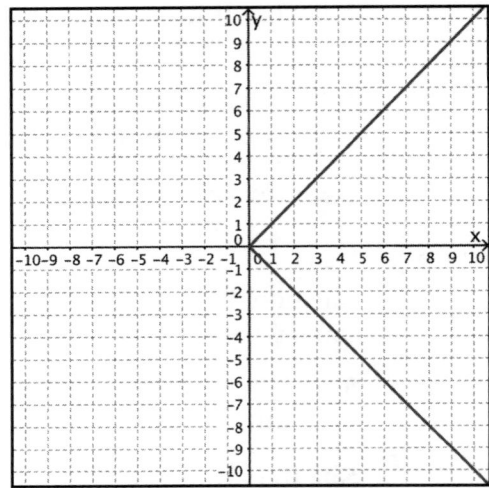

Explain why Moana is correct or incorrect in determining the graphed image to be a function.

10. Which of the following represents a relationship in which y is not a function of x?

A $y = 2$

B $x = -3$

C $x - 4y = -16$

D $y = -2(x - 6)^2 + 3$

EVALUATING FUNCTIONS

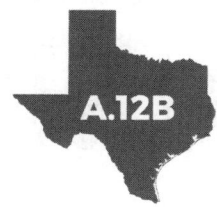

A.12B The student is expected to evaluate functions, expressed in function notation, given one or more elements in their domains.

ⓘ TELL ME MORE...

A **function** is a relationship between two variables: an **independent variable**, also called an input value, and a **dependent variable**, also called an output value. The values of the dependent variable are a function of the values of the independent variable. The **domain** of the function is the set of possible values for the independent variable. The **range** of the function is the set of possible values for the dependent variable.

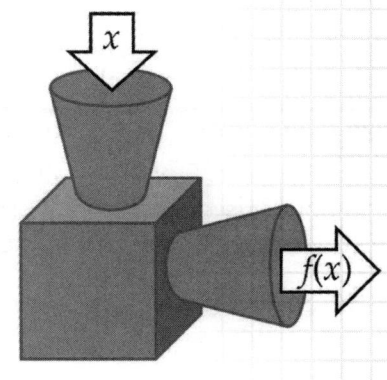

independent variable	dependent variable
domain input	range output

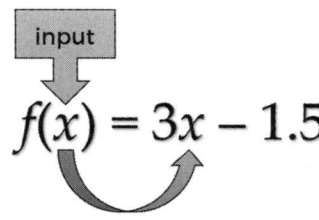

$$f(x) = 3x - 1.5$$

You can write a function using several different representations. **Function notation** uses the function name, in this case, f, and parentheses to indicate the independent variable or input values, in this case, x. It is read "f of x" as short for "f is a function of x." Be careful not to confuse these parentheses with those that indicate multiplication!

If you know the symbolic representation, written in function notation, then you can determine a specific function value if you know an input value.

✓ EXAMPLES

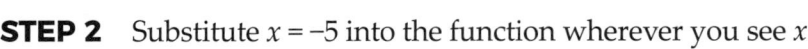

EXAMPLE 1: Given $k(x) = 1.5(3x + 7)$, what is the value of $k(-5)$?

STEP 1 Identify the domain value (value of the independent variable, x) that you will substitute into the function, $k(x)$.

$x = -5$

$k(-5)$ ← input

STEP 2 Substitute $x = -5$ into the function wherever you see x.

$$k(x) = 1.5(3x + 7)$$
$$k(-5) = 1.5(3(-5) + 7)$$

STEP 3 Use the order of operations to simplify the numerical expression.

$k(-5) = -12$

$$k(-5) = 1.5(3(-5) + 7)$$
$$k(-5) = 1.5(-15 + 7)$$
$$k(-5) = 1.5(-8)$$
$$k(-5) = -12$$

STEP 4 If the question is a gridded response question, enter your response on the grid provided. Practice using the grid with the instructions below.

1. Since the answer, −12, is an integer, you do not need to use a decimal point.
2. Record a negative in the space above the first column of the sign of the answer. Bubble the negative sign (−) in the first column.
3. Record a 1 in the first column containing numbers and a 2 in the next column.
4. Bubble the **1** beneath the numeral 1. Bubble the **2** beneath the numeral 2.

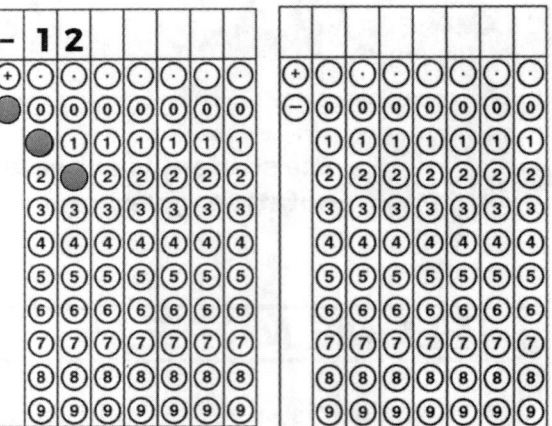

EXAMPLE 2: The Carancahua Trail High School choir sells yellow roses for Texas Independence Day as a fundraiser. Floral Supply uses the function $f(x) = 0.75x + 25$ to determine the cost of x roses. Flower Warehouse uses the function $g(x) = 1.15x$ to determine the cost of x roses. If the choir needs to order 200 roses, what is the cost difference between the two companies?

STEP 1 Determine the value of x that you will need to substitute into each function.

The number of roses is the independent variable.

The choir will need to order 200 roses, so $x = 200$.

$x = 200$

STEP 2 Substitute $x = 200$ into each function, $f(x)$ and $g(x)$. Use the order of operations to simplify each resulting expression.

$$f(x) = 0.75x + 25$$
$$f(200) = 0.75(200) + 25$$
$$f(200) = 150 + 25$$
$$f(200) = 175$$

$$g(x) = 1.15x$$
$$g(200) = 1.15(200)$$
$$g(200) = 230$$

$f(200) = 175$ and $g(200) = 230$

STEP 3 Determine the difference between $f(200)$ and $g(200)$. "Difference" means subtraction. The question simply asked for the cost difference and did not specify an order, so subtract the smaller number from the larger number.

$$\begin{array}{r} 230 \\ -\ 175 \\ \hline 55 \end{array}$$

55

> **YOU TRY IT!**
>
> Let $j(x) = 3x^2 - x$ and $m(x) = -4x + 3.2$. What is the sum of $j(4)$ and $m(4)$?
>
> $j(4) =$ _____
>
> $m(4) =$ _____
>
> $j(4) + m(4) =$ _____

EXAMPLE 3   A demographer is a person who studies population data and characteristics. Dr. Nguyen found that the population of Austin, Texas, follows the function $p(x) = 818.97(1.025)^x$, where x represents the number of years since 2010 and $p(x)$ is the population of Austin in thousands. According to her model, what will the population of Austin be in the year 2025?

STEP 1   Determine the value of x that you will need to substitute into the population function, $p(x)$.

The number of years since 2010 is the independent variable.

2025 − 2010 = 15, so 2025 is 15 years after 2010.

x = 15

STEP 2   Substitute $x = 200$ into $p(x)$.

$$p(x) = 818.97(1.025)^x$$
$$p(15) = 818.97(1.025)^{15}$$

STEP 3   Use the order of operations to simplify the numerical expression for $p(15)$.

$$p(15) = 818.97(1.025)^{15}$$
$$p(15) = 818.97(1.4483)$$
$$p(15) = 1186.11$$

p(15) = 1186.11, so the population of Austin in 2025 should be 1,186,110 people.

PRACTICE

1. Given that $f(x) = -(4 - 0.5x)$, find $f(-6)$.

3. Given $f(x) = 3x - 5$ and $h(x) = -\frac{3}{4}x + 7$, what is $f\left(\frac{2}{3}\right) + h(-18)$?

2. Jamie is studying abroad this summer and exchanges some money for her travels. The function $f(x) = 0.93x + 6.9$ describes the rate of exchange from U.S. dollars to the foreign currency plus a foreign transaction fee. How much of the foreign currency does Jamie receive for an exchange of f($350)?

4. If $f(x) = 4(6 - x)$ and $g(x) = \frac{1}{2}(4 - 2x)$, what is the product of $f(5)$ and $g(5)$?

5. A toy rocket is launched into the air and the height of the rocket, $h(x)$, in feet can be modeled by the function $h(x) = 192x - 16x^2$ where x represents the time in seconds since launch. What is the height of the rocket at $h(6)$?

9. Given $f(x) = x^2 + 4x + 1$ and $g(x) = (x - 3)^2 - 5$, which of the following is true?

   A $f(4) < g(-3)$

   B $f(-2) > g(0)$

   C $f(3) = g(5)$

   D $f(-5) + 1.5g(2) = 0$

6. The value of a vehicle depreciates with age. Chad bought a new motorcycle for $12,682. The function $m(x) = 12,682(0.9)^x$ describes the value of the motorcycle where x represents the number of years since Chad purchased the motorcycle. What is $f(4)$, the value of Chad's motorcycle after 4 years?

10. The half-life of caffeine is 6 hours meaning the amount of caffeine in a person's body is reduced by half every 6 hours. A 12-ounce cup of coffee contains about 142 mg of caffeine. The function $c(x) = 142\left(\frac{1}{2}\right)^x$ represents the amount of caffeine in a person's body based on the number of 6-hour intervals since drinking the coffee. How much caffeine remains in a person's body 24 hours after drinking a 12-ounce cup of coffee?

   F 284 mg

   G 0.0625 mg

   H 0.00000846 mg

   J 8.875 mg

7. If $f(x) = \frac{1}{2}(4)^x$, what is $f(6)$?

8. Let $g(x) = 3(x - 1)^2 + 4$ and $h(x) = 4(2)^x$. What is $\frac{2}{3}(g(-2) + h(3))$?

ARITHMETIC AND GEOMETRIC SEQUENCES

A.12C
A.12D

The student is expected to identify terms of arithmetic and geometric sequences when the sequences are given in function form using recursive processes.

The student is expected to write a formula for the nth term of arithmetic and geometric sequences, given the value of several of their terms.

ⓘ TELL ME MORE...

An arithmetic sequence is a list of numbers that has a *constant difference or constant* addend between every pair of two consecutive numbers in the list. A **geometric sequence** is a list of numbers that has a *constant ratio* or *constant multiplier* between every pair of two consecutive numbers in the list.

ARITHMETIC SEQUENCE	**GEOMETRIC SEQUENCE**
1.5, 5, 8.5, 12, 15.5, ...	1, 1.5, 2.25, 3.375, ...
+3.5 +3.5 +3.5 +3.5	×1.5 ×1.5 ×1.5
starting point: 1.5	starting point: 1
constant addend: 3.5	constant addend: 1.5
An arithmetic sequence is a linear function with a domain (term number) restricted to the set of whole numbers.	*A geometric sequence is an exponential functions with a domain (term number) restricted to the set of whole numbers.*

Sequences can be represented one of two ways. A sequence is represented **explicitly** if you have a function rule that allows you to directly calculate the nth term of the sequence. To generate the explicit rule, you can use the constant addend (or multiplier) to work backwards in the list to the 0th term, $f(0)$, and use that as your starting point. Or, you can use $(n-1)$ as your independent variable.

Term Number, n	1	2	3	4	5
Term, $f(n)$	1.5	5	8.5	12	15.5

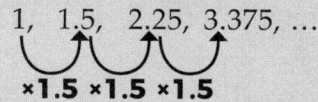 **$f(n) = 3.5(n – 1) + 1.5$**
If you know the term number, n, subtract 1 and then multiply by 3.5, the constant addend. Then add back the first term, 1.5.

A sequence is represented **recursively** if you calculate the nth term by adding to or multiplying by the previous $(n-1)$ term.

Term Number, n	1	2	3	4	5
Term, $f(n)$	1.5	5	8.5	12	15.5

$a_1 = 1.5; a_n = a_{n}-1 + 3.5$
The first term is 1.5. To get the next term, add 3.5, the constant addend, to the previous term. To get the nth term, an, take the previous term, $a_n - 1$, and add 3.5, the constant addend.

To use a recursive rule, you must know the term before the one you want to calculate. So if you want to calculate the 10th term of a sequence recursively, you must know the value of the 9th term. An advantage of using an explicit rule is that you can instantly calculate the nth term, like the 10th term, without knowing all of the previous terms. You can also represent the arithmetic sequence (1.5, 5, 8.5, 12, 15.5, …) using any of the following equivalent recursive forms.

$$a_1 = 1.5; a_n = a_{n-1} + 3.5$$
$$a_1 = 1.5; a_{n+1} = a_n + 3.5$$

$$f(1) = 1.5; f(n) = f(n-1) + 3.5$$
$$f(1) = 1.5; f(n+1) = f(n) + 3.5$$

✓ EXAMPLES

EXAMPLE 1: The first five terms in a sequence are shown below. Write an expression that can be used to find the nth term.

$$-1\frac{1}{2}, \frac{1}{2}, 2\frac{1}{2}, 4\frac{1}{2}, 6\frac{1}{2}, \dots$$

STEP 1 Determine if there is a constant difference or a constant multiplier.

$-1\frac{1}{2}, \frac{1}{2}, 2\frac{1}{2}, 4\frac{1}{2}, 6\frac{1}{2}$

The sequence has a constant difference (addend) of 2 and is arithmetic.

$\cdots +2 \; +2 \; +2 \; +2$

STEP 2 To write a recursive rule using $f(n)$, let the value of $f(1)$ be the first term. Then, use $f(n-1)$ as the previous term along with the constant difference, d, to write $f(n)$.

$$f(1) = -1\frac{1}{2}, d = 2$$

$f(1) = -1\frac{1}{2}$; $f(n) = f(n-1) + 2$

STEP 3 To write a recursive rule using a_n, let the value of a_1 be the first term. Then, use this as the previous term, a_{n-1}, along with the constant difference, d, to write a_n.

$$a_1 = -1\frac{1}{2}, d = 2$$

$a_1 = -1\frac{1}{2}$; $a_n = a_{n-1} + 2$

EXAMPLE 2: The half-life of ruthenium-106 is approximately 1 year, meaning that after 1 year, half of the amount of the substance has decayed and half of the amount of the substance remains. The sequence below shows the amount of a sample of 150 g of ruthenium-106 remaining after a sequence of years has passed.

$$150, 75, 37.5, 18.75, \dots$$

Write an expression that can be used to find the nth term in this sequence.

STEP 1 Determine if there is a constant difference or a constant multiplier.

$150, \quad 75, \quad 37.5, \quad 18.75, \dots$

$\times 0.5 \quad \times 0.5 \quad \times 0.5$

The sequence has a constant multiplier of 0.5 and is geometric.

YOU TRY IT!

The first five terms of a sequence are shown.

$$288, 72, 18, 4.5, 1.125, \dots$$

Is the sequence arithmetic or geometric? How can you tell?

_____

_____

_____

Constant Difference/Multiplier: ___

_____

Starting Point: _____

Sequence Expression: _____

STEP 2 To write a recursive rule using $f(n)$, let the value of $f(1)$ be the first term. Then, use this as the previous term, $f(n-1)$, along with the constant multiplier, r, to write $f(n)$.

$$f(1) = 150, r = 0.5$$

$f(1) = 150$; $f(n) = 0.5f(n-1)$

STEP 3 To write a recursive rule using a_n, let the value of a_1 be the first term. Then, use this as the previous term, a_{n-1}, along with the constant multiplier, r, to write a_n.

$$a_1 = 150, r = 0.5$$

$a_1 = 150$; $a_n = 0.5a_{n-1}$

EXAMPLE 3 Monelle wrote a sequence that can be generated by using $a_n = 2.5a_{(n-1)}$, where $a_1 = 24$ and n is a whole number greater than 1. What are the first five terms in Monelle's sequence?

STEP 1 Use the sequence formula and a_1 to determine a_2. In this case, $n = 2$.

$a_2 = 60$

$$a_n = 2.5a_{(n-1)}$$
$$a_2 = 2.5a_{(2-1)} = 2.5a_1$$
$$a_2 = 2.5(24) = 60$$

STEP 2 Use a_2 to determine the value of a_3, where $n = 3$.

$a_3 = 150$

$$a_n = 2.5a_{(n-1)}$$
$$a_3 = 2.5a_{(3-1)} = 2.5a_2$$
$$a_3 = 2.5(60) = 150$$

STEP 3 Continue the process for a_4 and a_5, the 4th and 5th terms, respectively, of Monelle's sequence.

$$a_n = 2.5a_{(n-1)} \qquad\qquad a_n = 2.5a_{(n-1)}$$
$$a_4 = 2.5a_{(4-1)} = 2.5a_3 \qquad a_5 = 2.5a_{(5-1)} = 2.5a_4$$
$$a_4 = 2.5(150) = 375 \qquad a_5 = 2.5(375) = 937.5$$

The first five terms are 24, 60, 150, 375, and 937.5.

PRACTICE

The first few terms of each sequence are shown below. Determine if the sequence is arithmetic or geometric and then write the recursive rule for the sequence.

1. $-7, -5, -3, -1, 1, \dots$

4. $10, 5.5, 1, -3.5, -8, \dots$

2. $6, 9, 13.5, 20.25, 30.375, \dots$

5. $-7, -3, 1, 5, 9, \dots$

3. $-2, -3, -4\frac{1}{2}, -6\frac{3}{4}, \dots$

6. $\frac{1}{4}, 1, 4, 16, 64, \dots$

7. A pattern of dots is shown in the figures.

If the pattern continues what explicit rule can be used to find the total number of dots in figure n?

8. Bree designed a pattern using graphics on her computer.

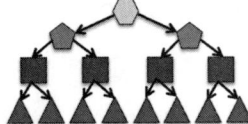

As the pattern moves from the top figure down each row of shapes, the number of shapes in each row forms a sequence. What formula describes this sequence?

9. Mr. Mahoney started a pecan orchard one autumn by planting 5 trees. Each autumn, Mr. Mahoney plants the same number of additional trees to extend his pecan orchard. The explicit sequence $f(n) = 5 + 11(n - 1)$ represents the number of trees in Mr. Mahoney's pecan orchard at the end of any number of years. How many trees will be planted in the orchard if the pattern continues for 42 years?

10. Samyra invested $250 buying shares of stock in her company. The sequence $a_0 = 250$, $a_n = 1.2a_{n-1}$ represents the amount of money Samyra's stock is worth at the end of n years. How much will Samyra's investment be worth at the end of year 5?

11. Mrs. Johannsen started a bacteria culture for a science experiment. The number of bacteria in the sample each hour follows the sequence $a_1 = 120$ and $a_n = 2a_{n-1}$. How many bacteria will be present at a_6?

A 796,262,400,000

B 7,680

C 3,840

D 152

12. Jamal is a long distance runner. While training for a marathon Jamal recorded the time spent running 5-mile tracks each week. The amount of time Jamal takes to run the tracks follows the sequence rule $a_1 = 50$ and $a_n = a_{n-1} - 0.30$ where n is the number of weeks training and the time to run a 5-mile track in week 1 was 50 minutes. If Jamal's training time continues to follow the sequence, how long should a 5-mile run take him after 7 weeks of training?

Record your answer and fill in the bubbles on your answer document.

SOLVING LITERAL EQUATIONS

A.12E The student is expected to solve mathematic and scientific formulas, and other literal equations, for a specified variable.

ⓘ TELL ME MORE...

Many fields, such as science, finance, and sports, use formulas to show how different quantities are related.

FORMULAS IN SCIENCE	FORMULAS IN FINANCE	FORMULAS IN SPORTS
Newton's Law of Gravitation	Simple Interest Formula	Fielding Percentage
$F = G\dfrac{m_1 m_2}{r^2}$	$I = Prt$	$F = \dfrac{A + P}{N}$
F = gravitational force	I = amount of interest	F = fielding percentage
m_1 and m_2 = masses of objects	P = principal investment	A = number of assists
r = distance between objects	r = interest rate	P = number of putouts
G is a constant	t = time	N = number of chances

You can use properties of algebra to solve any formula for any of the variables it contains. For example, in the simple interest formula, you can solve for P if you want to be able to calculate the amount of a principal investment required to generate a known amount of interest, I, in an account with a known interest rate, r, and time, t, of investment.

You may also encounter **literal equations**, where variables represent known values. The formula $d = rt$ that relates distance, rate, and time is an example of a formula that is also a literal equation. When you transform an equation from one form to another you are solving a literal equation.

✓ EXAMPLES

EXAMPLE 1: Write the linear function $5.5x - 2y = 24$ in slope-intercept form.

STEP 1 Slope-intercept form is $y = mx + b$. So, you will solve the mathematical formula, $5.5x - 2y = 24$ for y. Subtract $5.5x$ from both sides of the equation.

$$5.5x - 2y = 24$$
$$5.5x - 5.5x - 2y = -5.5x + 24$$

$$-2y = -5.5x + 24$$

STEP 2 Divide both sides of the equation by -2.

$$\frac{-2y}{-2} = \frac{-5.5x + 24}{-2} = \frac{-5.5}{-2}x + \frac{24}{-2}$$

$$y = 2.75x - 12$$

EXAMPLE 2: You can calculate the kinetic energy of an object with mass, m, and velocity, v, using the kinetic energy formula, $K = \frac{1}{2}mv^2$. Solve the kinetic energy formula for the mass of the object, m.

STEP 1 Solving the kinetic energy formula for m means that you will use properties of algebra to undo the operations that are being done to m. First, divide both sides of the equation by $\frac{1}{2}$ to un-multiply by $\frac{1}{2}$. (Note: dividing by $\frac{1}{2}$ is mathematically equivalent to multiplying by 2, which is the multiplicative inverse of $\frac{1}{2}$).

$$K = 2.5mv^2$$

$$K \div \frac{1}{2} = \frac{1}{2} \div \frac{1}{2} mv^2$$

2K = mv²

$$2K = mv^2$$

STEP 2 Un-multiply m by v^2 by dividing both sides of the equation by v^2.

$$\frac{2K}{v^2} = \frac{mv^2}{v^2}$$

$$\frac{2K}{v^2} = m$$

m = $\frac{2K}{v^2}$

EXAMPLE 3 An offensive ice hockey player's scoring percentage is calculated using the formula $S = \frac{A - G}{A}$, where S represents the scoring percentage, A represents the number of scoring attempts, and G represents the number of goals scored. Solve this formula for G?

STEP 1 Use the properties of algebra to undo the operations around G in the right member of the equation. Begin by un-dividing by the denominator, A. Multiply both sides of the equation by A.

$$S \times A = \frac{A - G}{A} \times A$$

SA = A – G

STEP 2 Un-add A from $- G$. Subtract A from both sides of the equation.

$$SA = A - G$$

$$SA - A = A - A - G$$

SA – A = –G

STEP 3 Un-multiply by −1 to make the sign of G positive. Divide all terms by −1.

$$\frac{SA - A}{-1} = \frac{-G}{-1}$$

$$-SA + A = G$$

G = –SA + A

YOU TRY IT! #1

The total cost of a car loan, L, can be calculated using the following formula.

$$L = Prt$$

In this formula, P is the amount of the loan, r is the annual interest rate, and t is the number of years of the car loan. Solve this formula for r.

$r = $ _____

YOU TRY IT! #2

The formula for the lateral surface area of a cylinder is $L = 2\pi rh$, where L is the lateral area, r is the radius of the base, and h is the height of the cylinder. Complete the steps below to solve the formula for r.

$$L = 2\pi rh$$

$$\frac{L}{\boxed{}} = \frac{2\pi rh}{\boxed{}}$$

$$\frac{L}{\boxed{}} = r$$

$$r = \frac{\boxed{}}{\boxed{}}$$

$r = $ _____

1. What is $5x - 4y = -24$ when written in terms of y, in slope-intercept form?

2. The formula $d = rt$ can be used to solve for distance when the average rate of speed and the time is known. What is the formula in terms of rate, when distance and time are known?

3. If $b + 2c = 4d$, write the equation for c in terms of b and d.

4. The formula for potential energy is $P = mgh$ where P is potential energy, m is mass, g is gravity, and h is height. What equation can be used to represent the value of m?

5. The volume of a cylinder is represented using the formula $V = \pi r^2 h$ where V is the volume, r is the radius of the base of the cylinder, and h is the height of the cylinder. How is the formula expressed when written in terms of π?

6. The standard form of a linear equation is $3x + 2y = -18$. What is the equation that represents the value of x when y is known?

7. The equation $y = mx + b$ represents a linear function in slope-intercept form. The value of the y-intercept in a certain function is -2. What expression represents the value of m in the equation?

8. The expression $\frac{2 + 3p}{q}$ is equivalent to the variable r. What expression is equivalent to the variable p?

9. The formula for the area of a trapezoid is represented as $A = \frac{x+y}{2}(h)$ where A is the area, x and y are the bases of the trapezoid, and h is the height of the trapezoid. Which of the following represents the equation in terms of x?

A $x = 2Ah - y$

B $x = \frac{2A}{hy}$

C $x = \frac{2Ah}{y}$

D $x = \frac{2A}{h} - y$

10. The formula for the perimeter of a rectangle is $P = 2l + 2w$ where P represents the perimeter, l is the length of the rectangle, and w is the width of the rectangle. What is the formula that can be used to find the length, l, of the rectangle?

F $l = \frac{2w - P}{2}$

G $l = \frac{P - 2w}{2}$

H $l = \frac{P - w}{2}$

J $l = P - w$

Unit 2
Describing and Graphing Linear Functions, Equations, and Inequalities

DETERMINING SLOPES

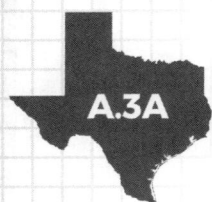

A.3A The student is expected to determine the slope of a line given a table of values, a graph, two points on the line, and an equation written in various forms, including $y = mx + b$, $Ax + By = C$, and $y - y_1 = m(x - x_1)$.

(i) TELL ME MORE...

The **rate of change** of a linear function describes how the dependent variable changes with respect to the independent variable. The graph of a linear function is a line, and the rate of change in this form can be called the slope of the line, which is a measure of the steepness of the line as a ratio of the vertical change to the horizontal change between any two points.

Although slope refers to the steepness of a line, you can calculate the slope of a line even if the representation you are given isn't a graph. A table represents a set of points that would appear on a line if you graphed them. Equations represent a relationship between two variables that would be linear if you graphed it.

You can use two points, an equation, a table, or a graph to determine the slope of the line represented in the given relationship.

SLOPE OF A LINE

In general, the slope of a line is the ratio of the vertical change to the horizontal change. If x is the horizontal axis and y is the vertical axis, then the slope, m, is $\frac{\Delta y}{\Delta x}$. The slope formula is:

$$m = \frac{\Delta y}{\Delta x} = \frac{y_2 - y_1}{x_2 - x_1}$$

✓ EXAMPLES

EXAMPLE 1: What is the slope of the line passing through the points $(-7, 8.5)$ and $(11, -3.5)$?

STEP 1 Calculate the slope using the slope formula. Let $(-7, 8.5) = (x_1, y_1)$ and $(11, -3.5) = (x_2, y_2)$.

$$m = \frac{y_2 - y_1}{x_2 - x_1} = \frac{-3.5 - 8.5}{11 - (-7)} = \frac{-12}{18} = -\frac{2}{3}$$

$$m = -\frac{2}{3}$$

EXAMPLE 2: The table represents some points on the graph of a linear function. What is the slope of the line represented by the points in the table?

STEP 1 A table contains sets of points that lie on the graph of a line. Select 2 points (pairs of x- and y-values).

(0, –7) and (6, –2)

x	y
–3	–9.5
0	–7
4	$-3\frac{2}{3}$
6	–2
7	$-1\frac{1}{6}$

STEP 2   Calculate Δx and Δy from the table.

x	y
−3	−9.5
0	−7
4	$-3\frac{2}{3}$
6	−2
7	$-1\frac{1}{6}$

$\Delta x = 6 - 0 = 6$ $\Delta y = -2 - (-7) = 5$

$$m = \frac{5}{6}$$

STEP 3   Determine the ratio, $m = \frac{\Delta y}{\Delta x}$, which is the slope of the line.

$$m = \frac{\Delta y}{\Delta x} = \frac{5}{6}$$

The slope of the line is $\frac{5}{6}$.

EXAMPLE 3   Line a is represented by the equation $-3x + 7y = 19$. Line b has the same slope as line a. What is the slope of line b?

STEP 1   The equation of line a is given in standard form, so rewrite it in slope-intercept form, $y = mx + b$.

$$-3x + 7y = 19$$
$$-3x + 3x + 7y = 19 + 3x$$
$$7y = 3x + 19$$
$$\frac{7y}{7} = \frac{3x + 19}{7}$$
$$y = \frac{3}{7}x + \frac{19}{7}$$

$$y = \frac{3}{7}x + \frac{19}{7}$$

STEP 2   Since the equation is in slope-intercept form, the coefficient of x represents the slope of line a.

$$y = \left(\frac{3}{7}\right)x + \frac{19}{7}$$

The slope of line a is $\frac{3}{7}$.

STEP 3   Line b has the same slope as line a.

The slope of line b is also $\frac{3}{7}$.

YOU TRY IT!

What is the slope of the line represented by the equation $y = 3.7x - 13.6$?

Is the equation in slope-intercept form? How can you tell?

_____

_____

What number represents m, the coefficient of x?_____

What is the slope of the line?

_____

EXAMPLE 4 What is the slope of line k?

STEP 1 Identify two points on the graph that appear to be integers.

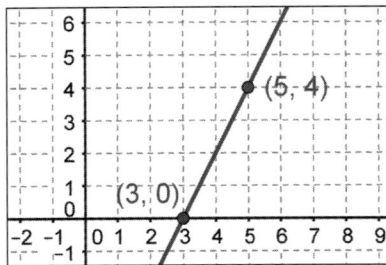

(3, 0) and (5, 4)

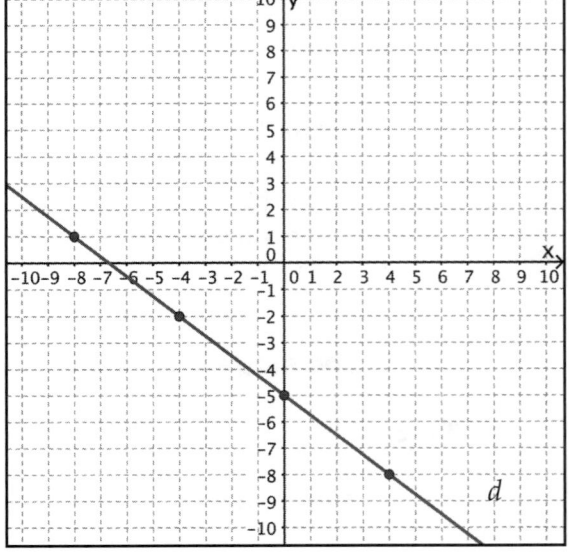

STEP 2 Count the slope in the graph using Δx, the change in x between the two points, and Δy, the change in y between the two points.

The slope of line k is 2

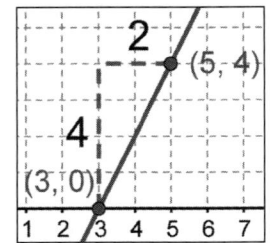

$$m = \frac{\Delta y}{\Delta x} = \frac{4}{2} = 2$$

PRACTICE

Answer the questions that follow.

1. A straight line passes through the points (−2, 6) and (5, −8). What is the slope of the line?

2. Jonah wrote the equation $1.2b + 4.8m = 48$ to represent the cost of hamburger buns and meat purchased for a group picnic. If Jonah's equation were graphed using b as the dependent variable, what would be the slope of the line formed by his equation?

3. The graph shows line d. What is the slope of line d?

4. The points in the table below lie along line k when graphed. What is the slope of line k?

x	y
-4	1
0	2
2	$2\frac{1}{2}$
5	$3\frac{1}{4}$
8	4

5. The graph of line p contains the following points.

x	y
-5	1
	7

If the slope of line p is $\frac{2}{5}$ what is the missing x-value from the table?

6. The graph of a line passes through the points $(-2, -7)$ and $(-8, 5)$. Which of the following correctly shows how to determine the slope of the line?

A $m = \dfrac{5+7}{-8+2} = \dfrac{12}{-6} = -2$

B $m = \dfrac{5-7}{-8-2} = \dfrac{-2}{-10} = \dfrac{1}{5}$

C $m = \dfrac{-8+2}{5+7} = \dfrac{-6}{12} = -\dfrac{1}{2}$

D $m = \dfrac{-8-2}{5-7} = \dfrac{-10}{-2} = 5$

7. Miriam was given the following table of values and asked to compute the slope of the line represented by the table.

x	0	15	22	30
y	18	21	22.4	24

Miriam determined the slope to be 5. What mistake did Miriam make in her computation?

8. Bruno and Li each wrote the equation of a line. Bruno's equation was $3x - 2y = 16$. Li's equation was $2x - 3y = 24$. If the slope of the line was $\frac{2}{3}$, who wrote the equation correctly?

9. Which equation shows the same slope as the points in the table?

x	y
-3	−4.5
0	−2
4	$1\frac{1}{3}$
6	3

F $y = \dfrac{15}{2}x - 2$

G $y = \dfrac{5}{6}x - 2$

H $y = \dfrac{3}{25}x - 2$

J $y = \dfrac{3}{2}x - 2$

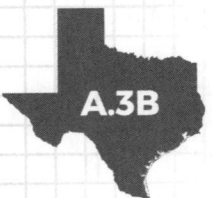

A.3B The student is expected to calculate the rate of change of a linear function represented tabularly, graphically, or algebraically in context of mathematical and real-world problems.

ⓘ TELL ME MORE...

The **rate of change** is a rate that shows how the dependent variable in a functional relationship changes with respect to the independent variable, or $\frac{\text{change in dependent variable}}{\text{change in independent variable}}$. In a linear function, the rate of change is constant. If the linear function is graphed, the rate of change is equivalent to the slope of the line that represents the linear relationship.

A rate is a comparison of two quantities with unlike units, so if the linear relationship is presented in a real-world context, it is important to include those units in the rate of change. For each situation below, record the rate of change from the graph, table, or equation.

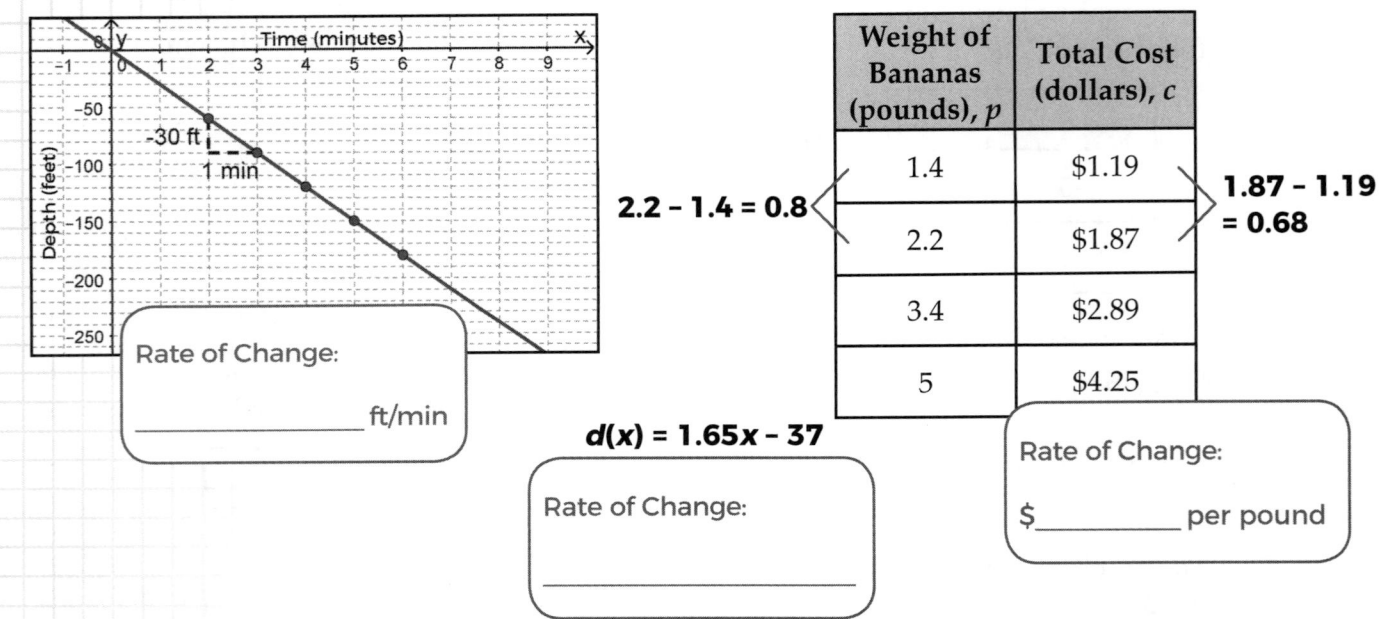

Rate of Change:

_____ ft/min

2.2 – 1.4 = 0.8

Weight of Bananas (pounds), p	Total Cost (dollars), c
1.4	$1.19
2.2	$1.87
3.4	$2.89
5	$4.25

1.87 – 1.19 = 0.68

Rate of Change:

$_____ per pound

$d(x) = 1.65x - 37$

Rate of Change:

_____

✓ EXAMPLES

EXAMPLE 1: The cost of bus rides, when purchased individually, for a local bus system creates a linear function. The table shows some costs for particular numbers of rides purchased. What is the rate of change of the total cost with respect to the number of bus rides?

STEP 1 Determine the change in the total cost and the change in the number of bus rides for two pairs of values in the table.

Number of Bus Rides, x	Total Cost (dollars), $c(x)$
5	$12.50
10	$25.00
15	$37.50
20	$50.00
25	$62.50

Number of Bus Rides, x	Total Cost (dollars), $c(x)$
5	$12.50
10	$25.00
15	$37.50
20	$50.00
25	$62.50

$10 - 5 = 5$ $25.00 - 12.50 = 12.50$

**The change in the total cost (dependent variable) is $12.50.
The change in the number of bus rides (independent variable) is
5 rides.**

STEP 2   Write the rate of change as a ratio. Simplify to a unit rate if possible.

$$\text{Rate of Change} = \frac{\text{change in dependent variable}}{\text{change in independent variable}} = \frac{\text{change in total cost}}{\text{change in number of bus rides}}$$

$$\text{Rate of Change} = \frac{\$12.50}{5 \text{ rides}} = \frac{\$2.50}{1 \text{ rides}} = \$2.50 \text{ per ride}$$

$2.50 per ride

EXAMPLE 2: The graph shows the temperature of an oven that is being preheated to 450°F over time. What is the rate of change in the temperature with respect to time?

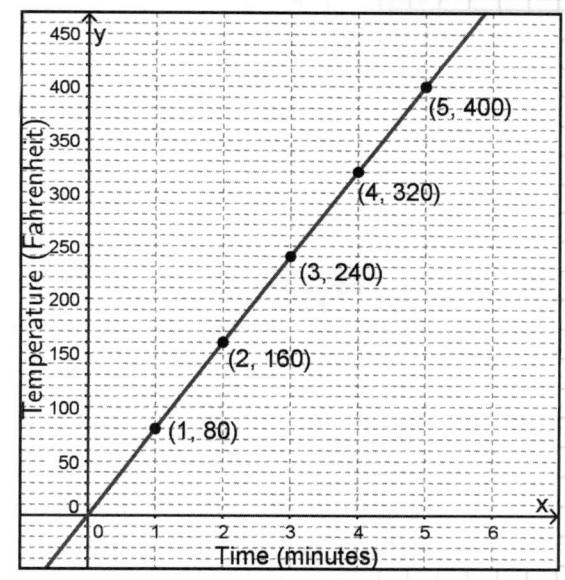

STEP 1   Determine the change in the dependent variable (vertical axis) between two known points.

   Change in temperature = 80°F

STEP 2   Determine the change in the independent variable (horizontal axis) between two known points.

   Change in time = 1 minute

STEP 3   Write the rate of change as the ratio of the change in the dependent variable to the change in the independent variable.

$$\text{Rate of Change} = \frac{\text{change in dependent variable}}{\text{change in independent variable}} = \frac{\text{change in temperature}}{\text{change in time}}$$

$$\text{Rate of Change} = \frac{80°F}{1 \text{ minute}} = 80°F \text{ per minute}$$

80°F per minute

EXAMPLE 3 Luz lives in a town that charges a $0.50 bag fee for paper bags if you do not bring your own reusable shopping bag. Luz purchases peaches at her local farmers' market. The equation $y = 1.25x + 0.50$ represents the cost, y, of purchasing x pounds of peaches including the $0.50 bag fee. What is the rate of change of the cost of peaches with respect to the number of pounds purchased?

STEP 1   Determine whether or not the equation is in slope-intercept form, $y = mx + b$. Compare the given equation with the general slope-intercept form.

$$y = mx + b$$
$$y = 1.25x + 0.50$$

The equation is in slope-intercept form.

STEP 2   Identify m, the coefficient of x, which will be the rate, $\frac{\text{change in dependent variable}}{\text{change in independent variable}}$.

$$y = \boxed{1.25}x + 0.50$$

m = 1.25

STEP 3   Interpret the rate of change to includes the units for the dependent variable and independent variable.

The dependent variable is cost in dollars. The independent variable is the amount of peaches in pounds.

$1.25 per pound

YOU TRY IT!

What is the rate of change of y with respect to x in the linear function

$2.5x - 5y = 50$?

Rewrite the equation in slope-intercept form.

What is the coefficient of x? _____

Rate of change: _____

PRACTICE

Answer the questions that follow.

1.   Alfonso released a helium balloon into the air and the recorded the height of the balloon in feet over time. The balloon's height can be represented by the linear function $f(x) = 800x + 6$ where x represents the time in minutes since release. What is the rate of change in the balloon's height with respect to time?

2.   The table shows the amount of money in Erika's savings account each month. What is the rate of change of the account balance, in dollars, with respect to the number of months Erika has been saving?

Month	Balance
0	$235
2	$339
5	$495
12	$859

3. The graph represents a linear function. What is the rate of change in y with respect to x for this function?

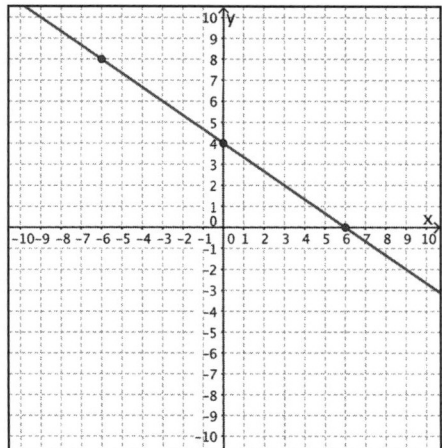

4. The graph shows Marla's cost of driving her vehicle. What is the meaning of the rate of change of cost with respect to distance in the graph of the line?

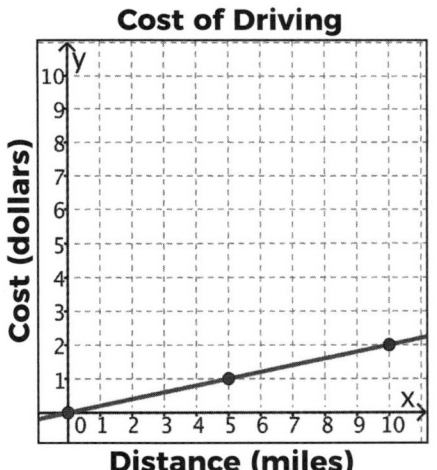

Cost of Driving

5. A scuba diver records his depth below the surface of the water. At 10 seconds he records a depth of 30 feet below sea level and at 40 seconds he records a depth of 100 feet below sea level. What is the diver's rate of change of depth with respect to time?

6. Miguel is on a biking trip with friends. On day one they ride 8 hours and travel 92 miles. On day two they ride 6 hours and travel 69 miles. On day 3 they ride 7 hours and travel 80.5 miles. What is the rate of change in miles with respect to hours spent riding? What is the meaning of the rate of change in the situation?

7. The graph below shows the distance-time relationship for two cars traveling on the same road. What is the difference in the rate of change for car A compared to car B?

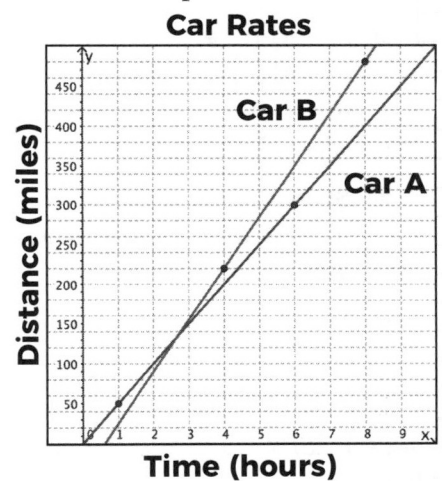

Car Rates

8. Lucy bought a new car for $35,000 in 2006. In 2011, Lucy's car was worth $15,500. If the car's value given its age is represented by a linear relationship, what is the average rate of change in value of Lucy's car with respect to the age of the car in years?

A   −$19,500

B   −$3,900

C   −$7,000

D   −$3,100

GRAPHING LINEAR FUNCTIONS

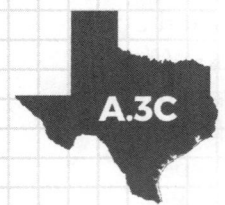

A.3C The student is expected to graph linear functions on the coordinate plane and identify key features, including *x*-intercept, *y*-intercept, zeros, and slope, in mathematical and real-world problems.

ⓘ TELL ME MORE...

The graph of a linear function is a visual representation of all of the ordered pairs, or paired numbers of input-output values, for that particular function. The constant rate of change makes the graph look like a line. Hence, the name *linear* function. You can graph a linear function as long as you know at least two points on the line. You can deduce information about those two points from an equation, verbal description, or table of values.

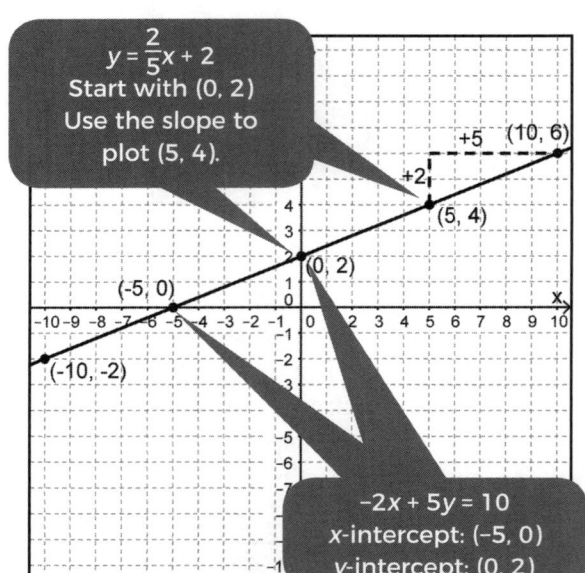

- **Slope-intercept form:** You can interpret the *y*-intercept from the symbolic form and use the slope to generate a second point.

- **Standard form:** You can use the symbolic form to determine the *x*-intercept and *y*-intercept.

- **Table of values:** Use any two paired values as two points on the graph.

- **Verbal description:** Look for information about a starting point or a pair of input-output values for the independent and dependent variables. Look for information about a rate of change and use that to generate a second point.

✓ EXAMPLES

EXAMPLE 1: Graph the equation $2x - 7y = 14$.

STEP 1 Determine the *x*-intercept, $(x, 0)$. Substitute $y = 0$ into the equation and solve for *x*.

x-intercept: (7, 0)

$$2x - 7y = 14$$
$$2x - 7(0) = 14$$
$$2x = 14$$
$$x = 7$$

STEP 2 Determine the *y*-intercept, $(0, y)$. Substitute $x = 0$ into the equation and solve for *y*.

y-intercept: (0, –2)

$$2x - 7y = 14$$
$$2(0) - 7y = 14$$
$$-7y = 14$$
$$y = -2$$

STEP 3 Plot the two points, x-intercept and y-intercept, and connect them with a line.

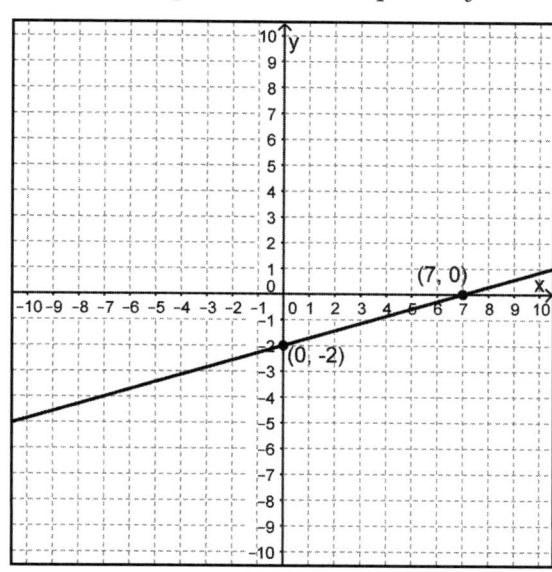

EXAMPLE 2: The table shows points contained on the line $2.5x - 2y = 10$. Graph the linear function represented by the points in the table.

x	-4	0	4	8
y	-10	-5	0	5

STEP 1 Select two paired values (ordered pairs) from the table. Write them as ordered pairs, (x, y).

x	-4	0	4	8
y	-10	-5	0	5

(4, 0)   (8, 5)

(4, 0) and (8, 5)

STEP 2 Graph the two points. Connect them with a line.

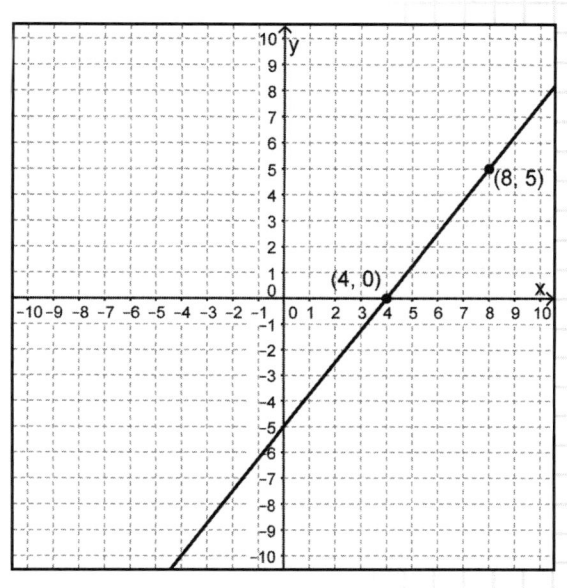

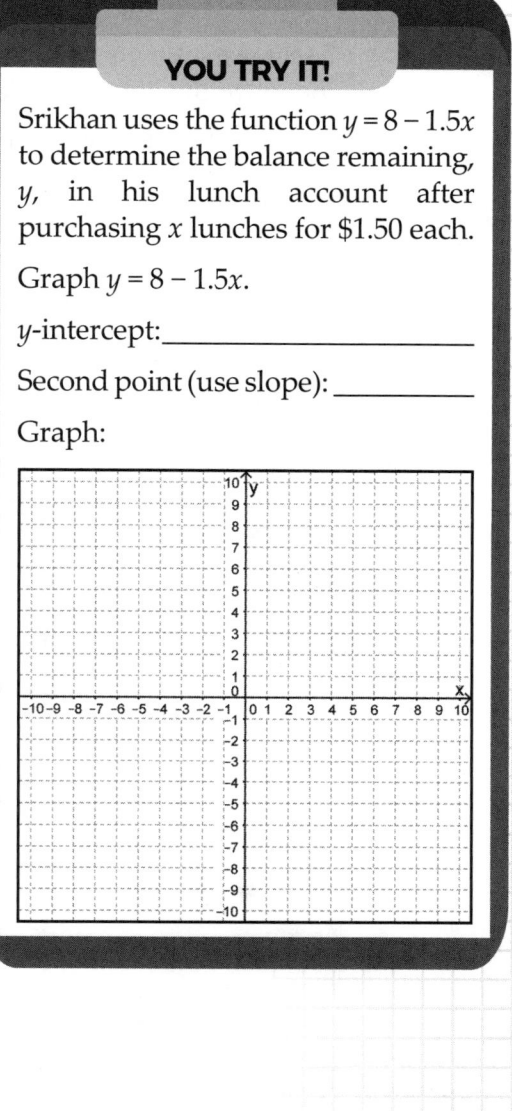

YOU TRY IT!

Srikhan uses the function $y = 8 - 1.5x$ to determine the balance remaining, y, in his lunch account after purchasing x lunches for $1.50 each.

Graph $y = 8 - 1.5x$.

y-intercept:_____

Second point (use slope): _____

Graph:

EXAMPLE 3 Sam earns $9.00 an hour working at a music store. If he works more than 30 hours a week he also earns a bonus of $3.00 more per hour. Company policy limits an employee to working no more than 50 hours each week. Graph the linear function that represents Sam's weekly earnings in dollars for working x hours over 30 hours.

STEP 1 Determine how much money Sam will make if he works 30 hours. This will be the starting point for the linear function for Sam's weekly earnings beyond 30 hours.

Number of Hours × Hourly Rate = Weekly Earnings for 30 Hours

30 × $9.00 = $270

$270

STEP 2 Identify the rate of change, or slope of the line.

Sam earns $9.00 + $3.00 = $12.00 per hour

$12 per hour

STEP 3 Graph the starting point, (0, 270), which is the y-intercept. Use the slope to generate a second point. If Sam worked 35 hours one week, he would earn $y = 270 + 5(12) = 330$ dollars. So a second point on the graph would be (5, 330).

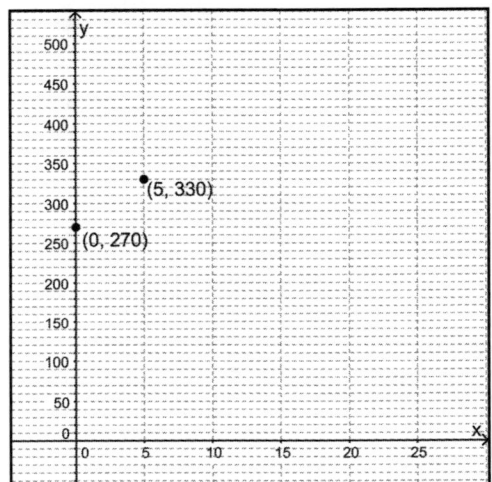

STEP 4 Connect the two points. Since Sam can work no more than 50 hours, there is an endpoint at 50 − 30 = 20, since x represents the number of hours worked *past* 30.

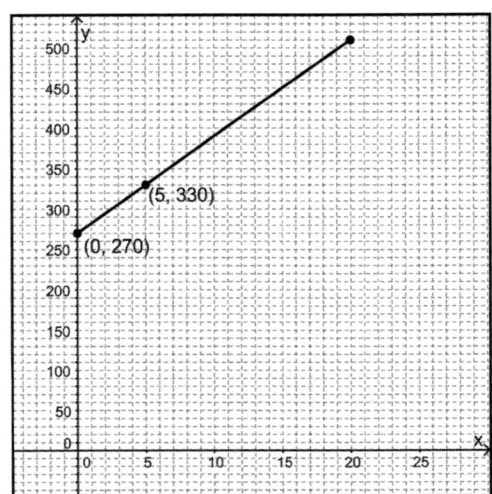

PRACTICE

Graph each of the following linear functions.

1. $y = \frac{1}{3}x - 5$

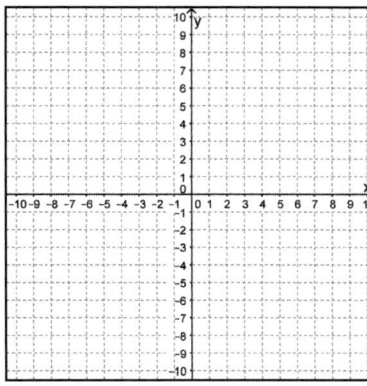

2. $y = -4x + 7$

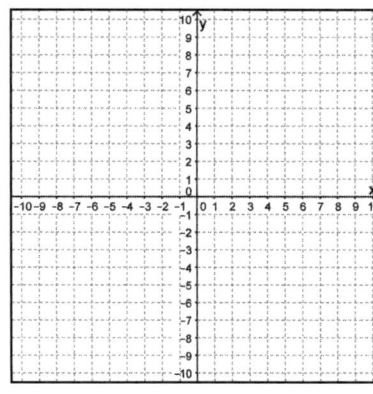

3. $4x - 6y = 12$

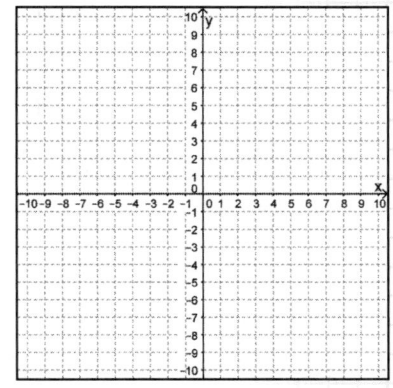

4. $x + 9y = 9$

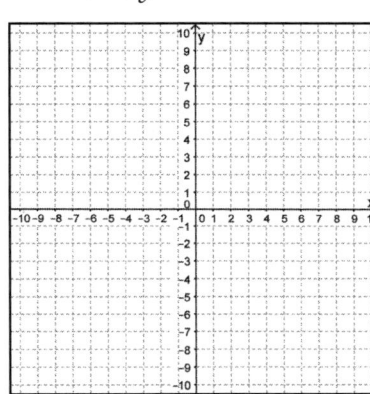

5.

x	-4	0	4	8
y	8	6	4	2

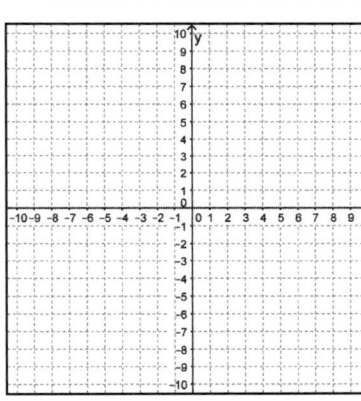

6.

x	0	1	4	5
y	-8	-5	4	7

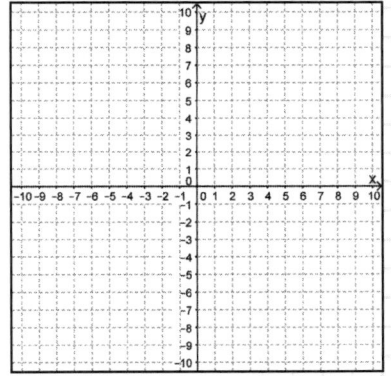

7. Jack is training for a bike race with his sister Tia. He rides at an average speed of 5 miles per hour and gives Tia a 4-mile head start because her speed is slower. The distance he travels, y, in miles versus time spent biking, x, in hours compared to Tia is represented by a linear function. Plot the graph of the linear function that represents Jack's bike training trip.

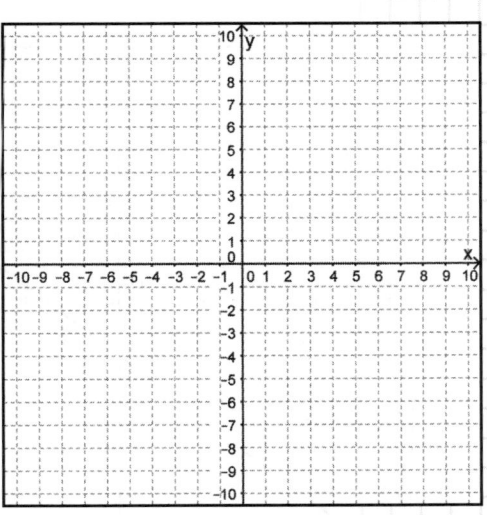

8. When line p is graphed its rate of change is $\frac{5}{6}$ and it has a zero of 12. Plot line p on the graph and give the equation of line p in standard form.

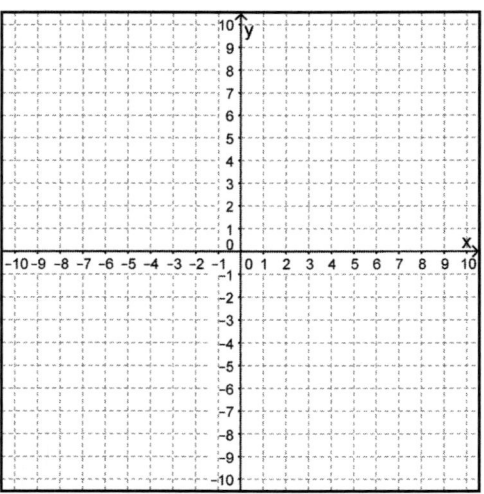

9. Which graph represents $-6x + 8y = 24$?

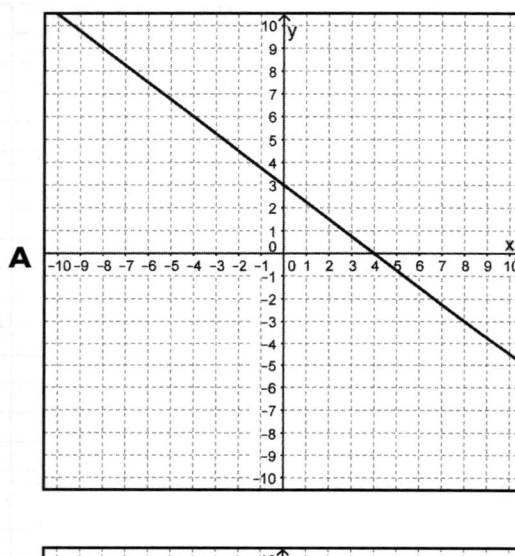

A

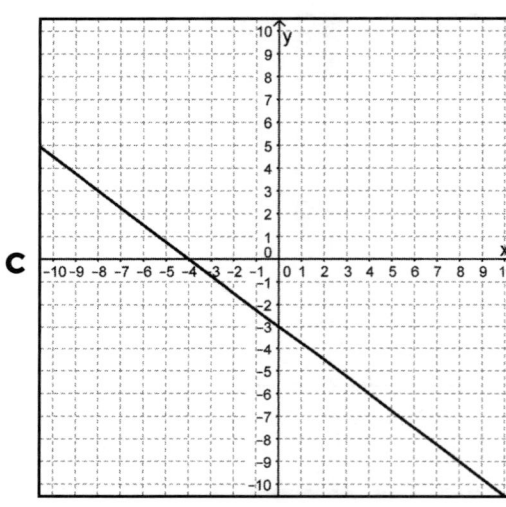

C

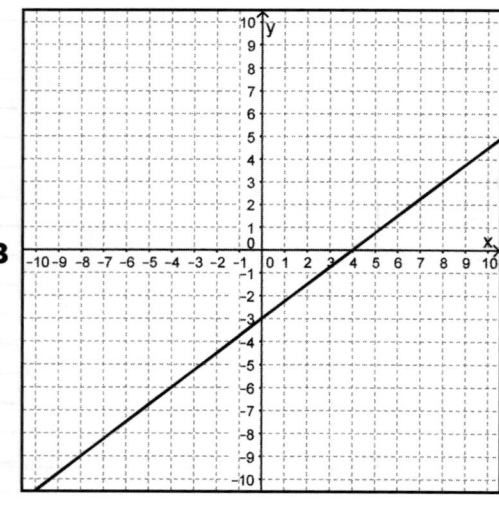

B

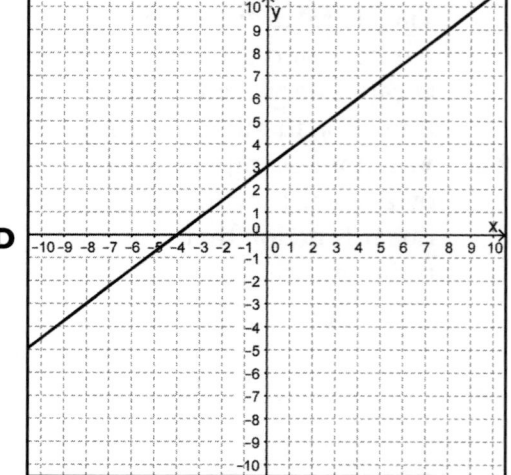

D

KEY FEATURES OF LINEAR FUNCTIONS

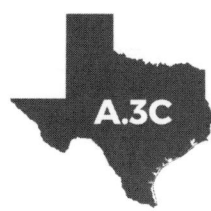

A.3C The student is expected to graph linear functions on the coordinate plane and identify key features, including *x*-intercept, *y*-intercept, zeros, and slope, in mathematical and real-world problems.

ⓘ TELL ME MORE...

The graph of a linear function reveals certain attributes that are important to the function, especially when you are using a function to model real-world data or real-world situations. For example, the graph of $y = \frac{2}{5}x + 2$ (or, $-2x + 5y = 10$) is shown. Key features of the graph include the *x*-intercept, *y*-intercept, zero, and slope.

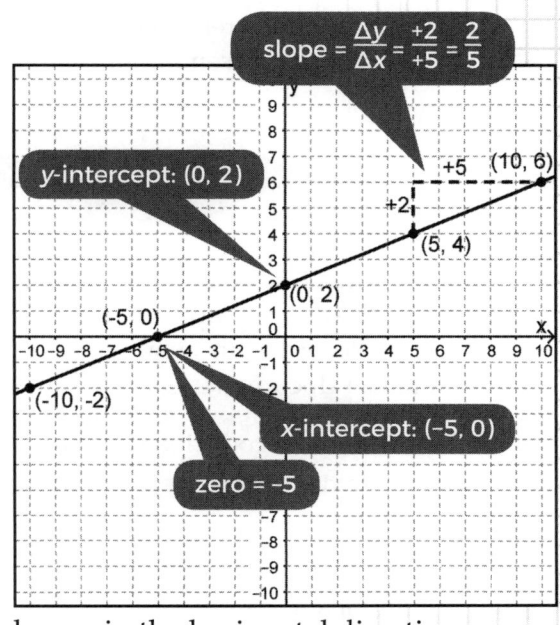

- The **x-intercept** is the point where the graph of the line crosses the *x*-axis. Here, the *y*-value is 0.

- The **y-intercept** is the point where the graph of the line crosses the *y*-axis. Here, the *x*-value is 0.

- The **zero** of a linear function is the input value that generates an output value of 0. It is equivalent to the *x*-coordinate of the *x*-intercept.

- The **slope** of a linear function is the steepness of the graph of the line. Slope is measured as the ratio of the change in the vertical direction to the change in the horizontal direction.

✓ EXAMPLES

EXAMPLE 1: The graph of a linear function is shown. What ordered pairs best represent the *x*-intercept and *y*-intercept of the line?

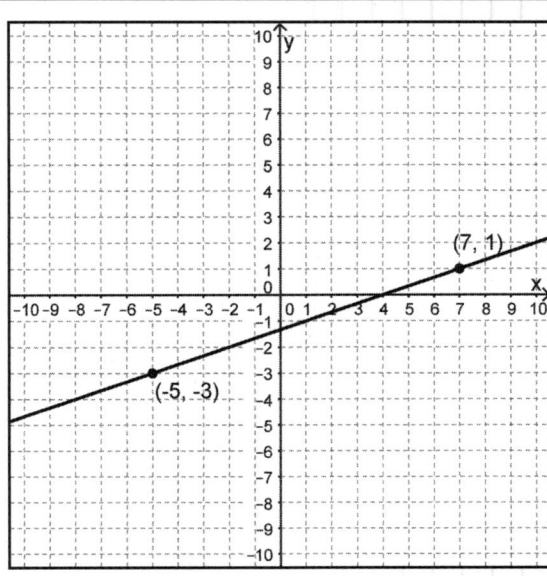

STEP 1 Use the coordinates of the two points provided to determine the slope of the line.

$$m = \frac{y_2 - y_1}{x_2 - x_1} = \frac{1 - (-3)}{7 - (-5)} = \frac{4}{12} = \frac{1}{3}$$

$$\boldsymbol{m = \frac{1}{3}}$$

STEP 2 Use the point-slope formula to determine the equation of the line shown.

$$\boldsymbol{y = \frac{1}{3}x - \frac{4}{3}}$$

$$y - y_1 = m(x - x_1)$$
$$y - 1 = \frac{1}{3}(x - 7)$$
$$y - 1 = \frac{1}{3}x - \frac{7}{3}$$
$$y - 1 + 1 = \frac{1}{3}x - \frac{7}{3} + \frac{3}{3}$$
$$y = \frac{1}{3}x - \frac{4}{3}$$

STEP 3   Determine the coordinates of the x-intercept, $(x, 0)$, by substituting $y = 0$ and solving for x.

(4, 0)

$$y = \frac{1}{3}x - \frac{4}{3}$$
$$0 = \frac{1}{3}x - \frac{4}{3}$$
$$0 = x - 4$$
$$4 = x$$

STEP 4   Determine the coordinates of the y-intercept, $(0, y)$, by substituting $x = 0$ and solving for y.

$$\left(0, -\frac{4}{3}\right)$$

$$y = \frac{1}{3}x - \frac{4}{3}$$
$$y = \frac{1}{3}(0) - \frac{4}{3}$$
$$y = (0) - \frac{4}{3}$$
$$y = -\frac{4}{3}$$

STEP 5   Check the reasonableness of your calculated intercepts by making sure the given graph crosses the line at these points.

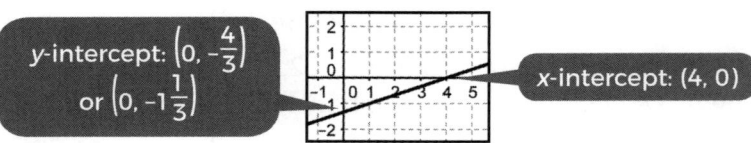

y-intercept: $\left(0, -\frac{4}{3}\right)$ or $\left(0, -1\frac{1}{3}\right)$ x-intercept: $(4, 0)$

EXAMPLE 2: The graph of $r(x)$ is shown. What is the zero of r? Record your answer and fill in the bubbles on your answer document.

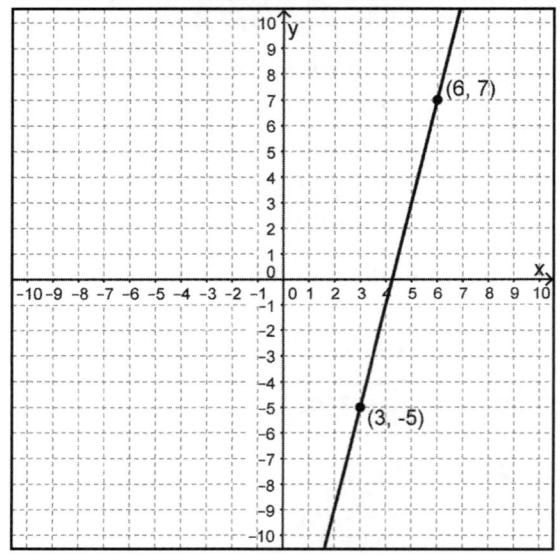

STEP 1   Use the coordinates of the two points provided to determine the slope of the line.

$$m = \frac{y_2 - y_1}{x_2 - x_1} = \frac{7 - (\text{-}5)}{6 - 3} = \frac{12}{3} = 4$$

$m = 4$

STEP 2   Use the point-slope formula to determine the equation of the line shown.

$$y - y_1 = m(x - x_1)$$
$$y - 7 = 4(x - 6)$$
$$y - 7 = 4x - 24$$
$$y = 4x - 17$$

$r(x) = 4x - 17$

STEP 3   The zero of r is the x-value that makes $r(x) = 0$. Substitute 0 for $r(x)$ and solve for x.

$$r(x) = 4x - 17$$
$$0 = 4x - 17$$
$$17 = 4x$$
$$\frac{17}{4} = x$$

The zero is $\frac{17}{4}$.

STEP 4 If the question is a gridded response question, enter your response on the grid provided. Practice using the grid with the instructions.

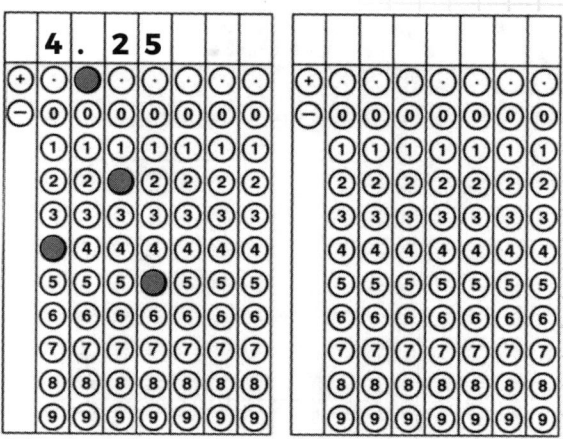

1. Since the answer, $\frac{17}{4}$, is a fraction, convert it to a decimal, 4.25.

2. Record a 4 in the first column containing numbers. Record a decimal point in the next column. Record and a 2 in the next column and a 5 in the next column.

3. Bubble the 4 beneath the numeral 4. Bubble the . beneath the decimal point. Bubble the 2 beneath the numeral 2. Bubble the 5 beneath the numeral 5

EXAMPLE 3 The graph shows the volume of propane, $V(x)$, in a propane tank after x months. What do the x-intercept, y-intercept, and slope of the graph represent.

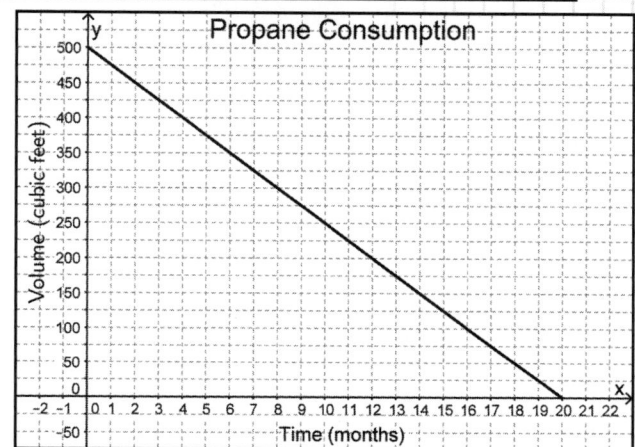

STEP 1 Identify the coordinates of the x-intercept from the graph, if possible.

(20, 0)

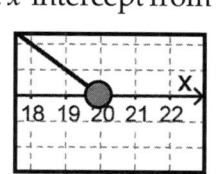

STEP 2 Interpret the coordinates of the x-intercept using the axis labels.
(20, 0) indicates 20 months and a volume of 0 cubic feet.

After 20 months, the volume of the propane tank was 0 cubic feet and the tank was empty.

STEP 3 Identify the coordinates of the y-intercept from the graph, if possible.

(0, 500)

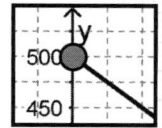

STEP 4 Interpret the coordinates of the x-intercept using the axis labels.
(0, 500) indicates 0 months and a volume of 500 cubic feet

In the beginning (0 months), the tank had 500 cubic feet of propane.

STEP 5 Identify the slope from the graph.

m **= –25 cubic feet per month**

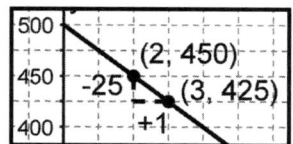

STEP 6 Interpret the slope using the axis labels.
A slope of –25 cubic feet per month indicates a decrease of 25 cubic feet of propane from the tank each month.

Each month, 25 cubic feet of propane was consumed.

PRACTICE

For each graph shown, identify the slope, x-intercept, y-intercept, and zero (if they exist).

1.

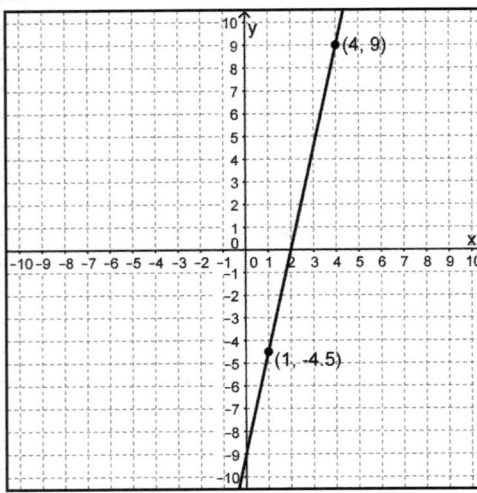

3.

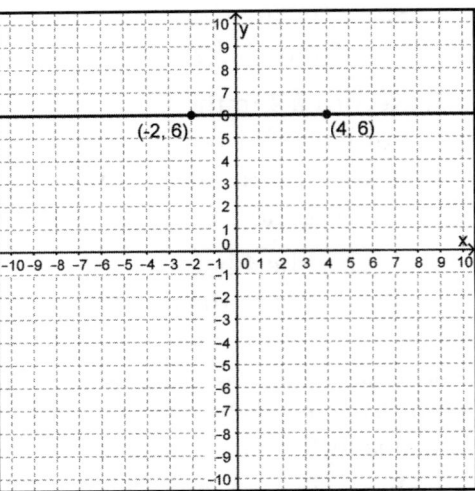

2.

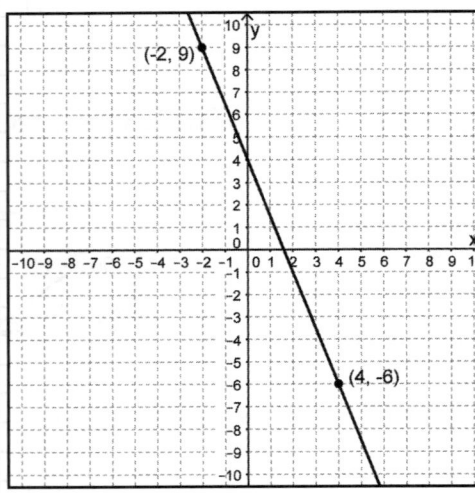

4.

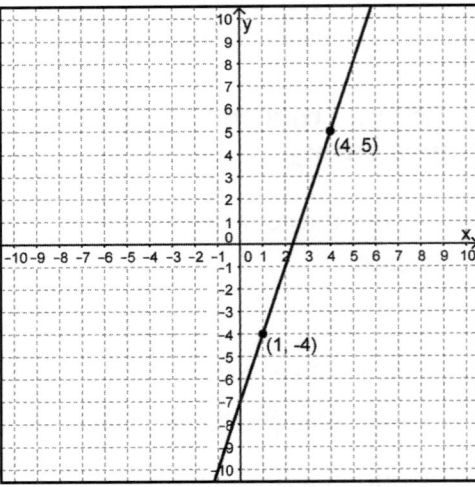

5. Petra runs a snow cone stand in town. She sells snow cones in two sizes. Small cones sell for $3 each and large cones sell for $5 each. On Thursday evenings Petra sells an average of $60 in snow cones. When the relationship of number of large cones sold, x, and the number of small cones sold, y, is plotted, what do the x- and y-intercepts represent?

6. The graph of linear function $h(x)$ is shown. What is the zero of $h(x)$? Record your answer and fill in the bubbles on your answer document.

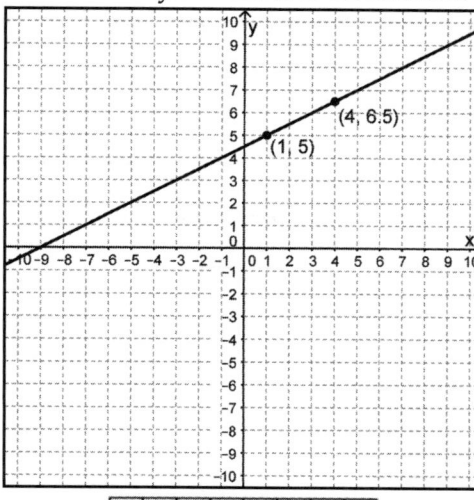

7. Maya's Coffee Club gift card balance information is shown in the table below.

Days used, x	Balance ($), y
13	71
14	68
20	50
26	32

If Maya writes a function to model the data on her gift card and graphs the function, where will the x- and y-intercepts be located?

8. Carolina often calls her mother who lives in Canada. A 6-minute international call to her mother costs $7 and a 15-minute international call costs $10. On the graph of the function $f(m)$ that represents the situation, what do the slope and y-intercept represent?

A The slope is the connection fee of $5 and the y-intercept is the cost per minute.

B The slope is the cost per minute and the y-intercept is the cost of a 5-minute call.

C The slope is the cost per minute and the y-intercept is the connection fee of $5.

D The slope is the cost to connect the call each minute and the y-intercept is the cost after talking 15-minutes.

9. The graph of a linear function is shown below.

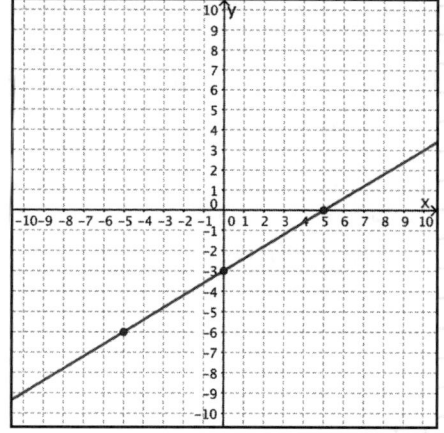

How will the graph change if the slope remains the same and the zero is changed to 7?

F The graph will shift up 10 units.

G The graph will shift down 2 units.

H The graph will get steeper.

J The graph will be less steep.

GRAPHING LINEAR INEQUALITIES (TWO VARIABLES)

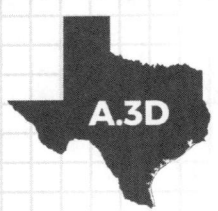

A.3D | The student is expected to graph the solution set of linear inequalities in two variables on the coordinate plane.

ⓘ TELL ME MORE...

An **inequality** shows a relationship between two quantities that are not equivalent or equal. In this case, one quantity is less than the other. You can use the symbols for **less than**, <, or **greater than**, >, to show these relationships.

The graph shows the inequality $y < 2x + 3$. The **solution set** of this inequality is the set of all ordered pairs (x, y) that when substituted into the inequality make a true statement.

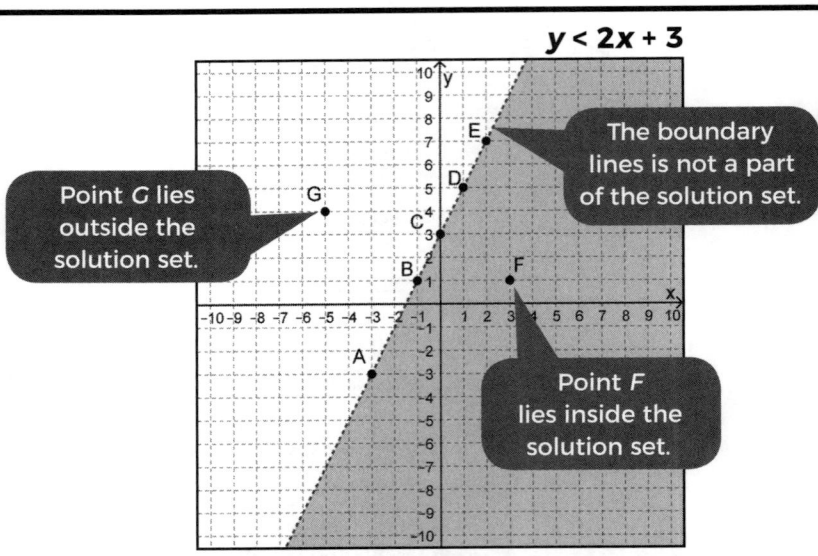

$y < 2x + 3$

The boundary lines is not a part of the solution set.

Point G lies outside the solution set.

Point F lies inside the solution set.

Points A, B, C, D, and E all lie on the **boundary line** of the inequality. Since the inequality reads "less than" and does not include "or equal to," the boundary line, including these five points, is not a part of the solution set.

Point A: $(-3, -3)$
$$-3 < 2(-3) + 3$$
$$-3 < -3$$
This is not a true statement, since $-3 = -3$.

Point F, $(3, 1)$ along with all other points in the shaded region, is a part of the solution set.
$$3 < 2(1) + 3$$
$$3 < 5, \text{ which is a true statement.}$$

Point G, $(-5, 4)$, along with all other points in the *unshaded region*, is *not* a part of the solution set.
$$4 < 2(-5) + 3$$
$$4 < -7$$
This not is a true statement, since $4 > -7$.

In a graph, a solid boundary line indicates that the points on the line are included in the solution set. Since this is the graph of a line, then the inequality will include "or equal to:" ≤ or ≥. If the boundary line is dashed, then those points are not included in the solution set and the inequality will be < or >.

✓ EXAMPLES

EXAMPLE 1: Graph the inequality $4x - 7y < 14$.

STEP 1 Since the inequality is in standard form, determine the x-intercept and y-intercept of the boundary line.
(3.5, 0) and (0, -2)

For the x-intercept, $(x, 0)$, let $y = 0$.
$$4x - 7y = 14$$
$$4x - 7(0) = 14$$
$$4x = 14$$
$$x = 3.5$$

For the y-intercept, $(0, y)$, let $x = 0$.
$$4x - 7y = 14$$
$$4(0) - 7y = 14$$
$$-7y = 14$$
$$y = -2$$

STEP 2 Plot the *x*- and *y*-intercepts on a graph. Connect them with a boundary line. The boundary line will be solid if the inequality includes "or equal to" or dashed if the inequality does not include "or equal to."

The inequality is <, or "less than." The inequality does not contain "or equal to" so the boundary line will be dashed.

$$4x - 7y \bigcirc 14$$

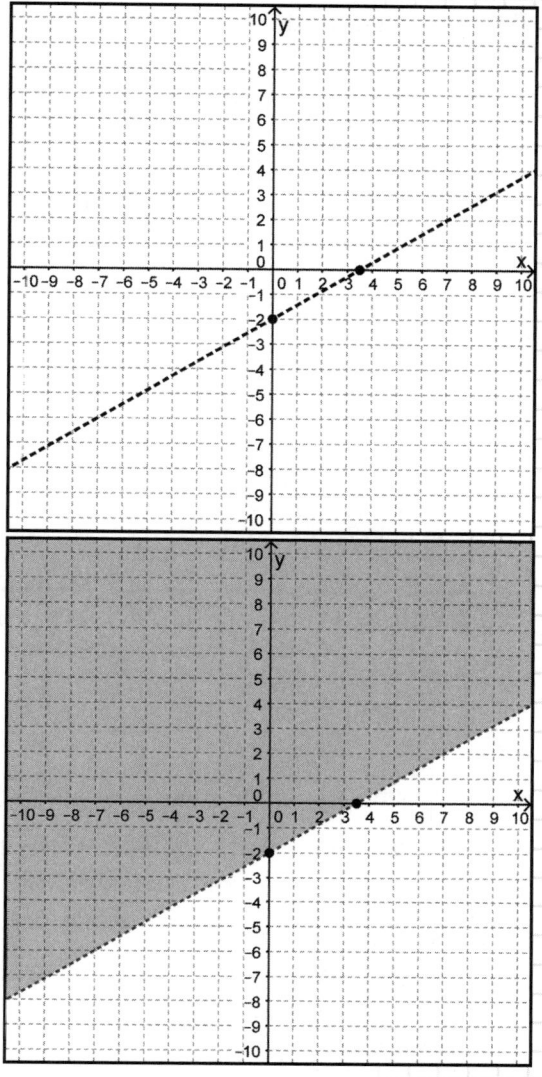

STEP 3 Determine which region to shade by selecting a test point, (0, 0). If the inequality is true, then shade the region that includes (0, 0). If not, then shade the other side of the boundary line.

$$4x - 7y < 14$$

$$4(0) - 7(0) < 14 \ ?$$

$$0 < 14$$

This inequality is true, so the solution set contains (0, 0) and shade the region above the line

---

EXAMPLE 2: Graph the inequality $y \leq -4x - 5$.

STEP 1 Since the inequality is in slope-intercept form, identify the *y*-intercept from *b* and use the slope to generate a second point.

$$b = -5, \text{ so the y-intercept is } (0, -5)$$

The slope, *m*, is $-4 = \dfrac{-4}{1}$, so $\Delta y = -4$ and $\Delta x = 1$.

y-intercept = (0, –5) and second point = (1, –9)

STEP 2 Plot the two points on a graph. Connect them with a boundary line. The boundary line will be solid if the inequality includes "or equal to" or dashed if the inequality does not include "or equal to."

The inequality is ≤, or "less than **or equal to**," so the boundary line will be solid.

$$y \bigcirc -4x - 5$$

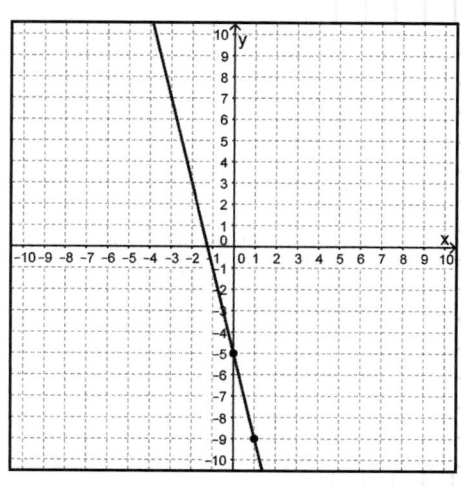

STEP 3 Determine which region to shade by selecting a test point, (0, 0). If the inequality is true, then shade the region that includes (0, 0). If not, then shade the other side of the boundary line.

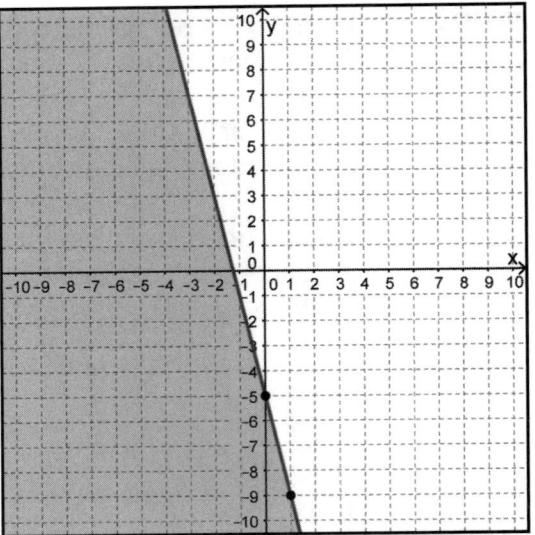

$$y \leq -4x - 5$$

$$0 \leq -4(0) - 5 \text{ ?}$$

$$0 \leq 0 - 5 \text{ ?}$$

$$0 \leq -5 \text{ ?}$$

This inequality is not true, so the solution set does not contain (0, 0). Shade the region on the other side of the boundary line.

EXAMPLE 3 Which of the following ordered pair(s) is in the solution set of $y \geq \frac{3}{7}x + 3$?

$$(-7, 0) \quad (0, 9) \quad (-4, 1) \quad (2, -5).$$

STEP 1 Graph the inequality $y \geq \frac{3}{7}x + 3$.

■ The inequality is in slope-intercept form, so plot the y-intercept and use the slope to generate an additional point.

■ The boundary line will be solid since the inequality contains "or equal to."

■ Test (0, 0). The resulting inequality, $0 \geq 3$, is not true, so shade the other side of the boundary line as the solution set.

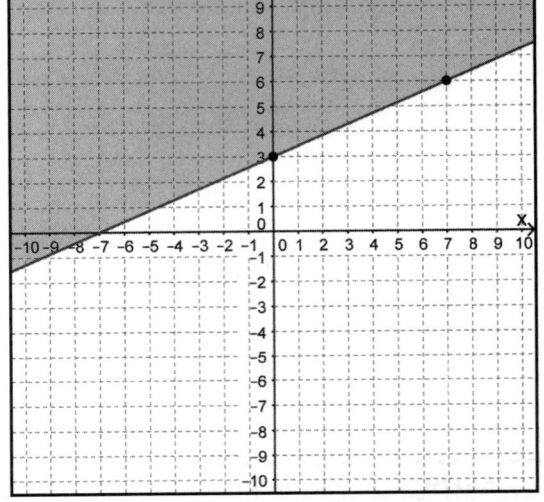

STEP 2 Plot the four given points. If the point lies in the shaded region or on the boundary line, then it is in the solution set of the inequality.

(–7, 0) and (0, 9) are in the solution set.

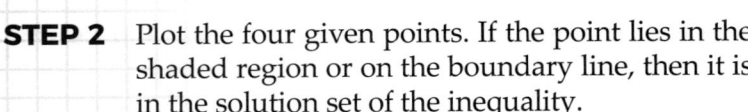

MAKE A NOTE . . .

Why are points on the boundary line a part of the solution set for some inequalities but not others? Record your thoughts here.

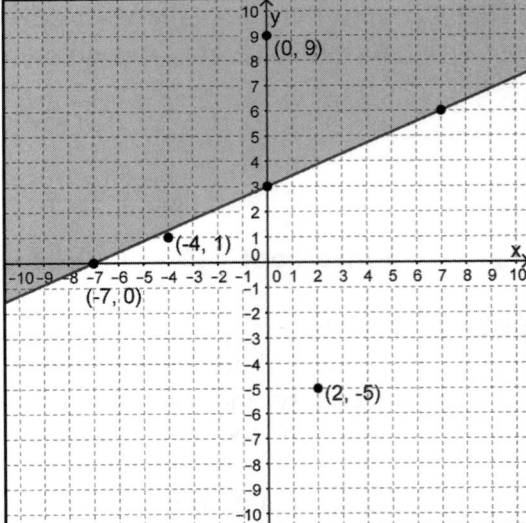

1. Shade the graph below to represent the solution set to the inequality $4x - 3y > -3$

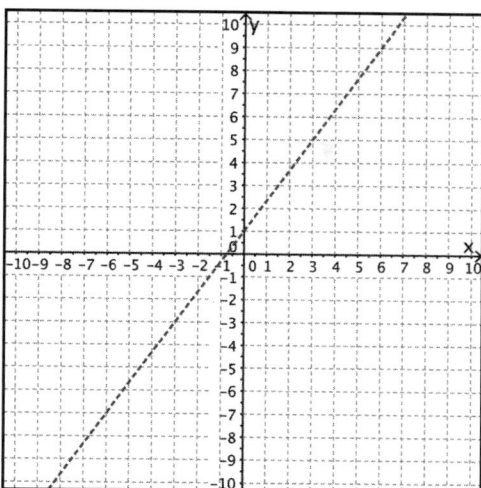

2. Graph the solution set for the linear inequality $3x + 2y < 14$

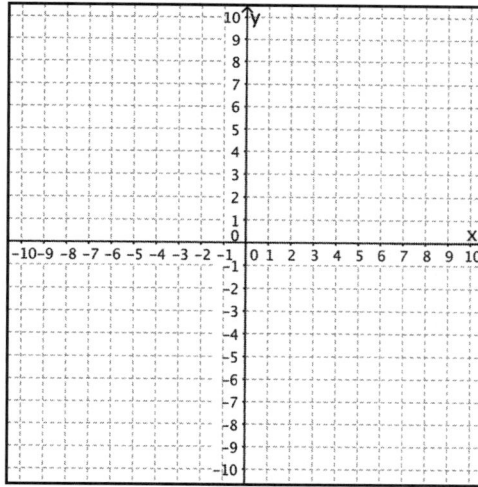

3. The points (5, −4) and (−4, −20) both lie in the solution set of the inequality shown. Which symbol belongs in the blank of the inequality statement to make this true?

$$y \underset{?}{\quad\rule{1.5em}{0.4pt}\quad} \frac{5}{4}x - 15$$

4. The inequality $200x + 125y < 2,500$ describes the number of average weight men, x, and women, y, who can ride on an elevator that must have a total weight of less than 2,500 pounds. Graph the possible solution set of the inequality.

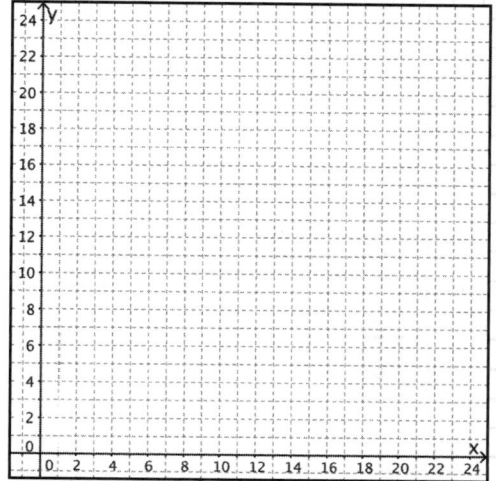

5. Carina plots a linear inequality such that the points (10, −4) and (−5, 14) are in the solution set and lie along the boundary line of the inequality. The origin also lies in the solution set. What linear inequality does Carina plot?

6. When the linear inequality $x < 0$ is plotted on the coordinate plane, which quadrants of the plane are shaded?

7. When the linear inequality $y > \frac{2}{5}x - 3$ is graphed, which of the following points lie in the solution set?

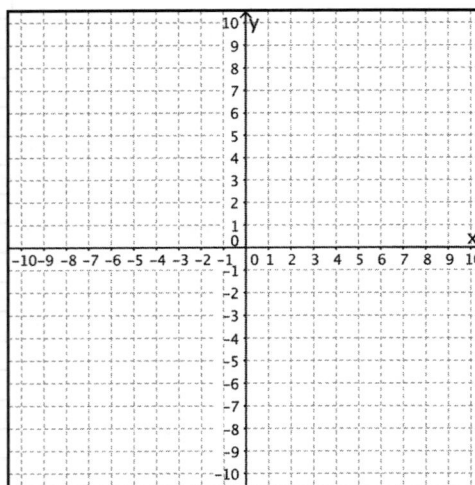

Point	A	B	C	D
x	15	0	4	-1
y	0	15	-6	3

8. Which of the following ordered pairs does NOT lie in the solution set of $3x - 4y < 18$?

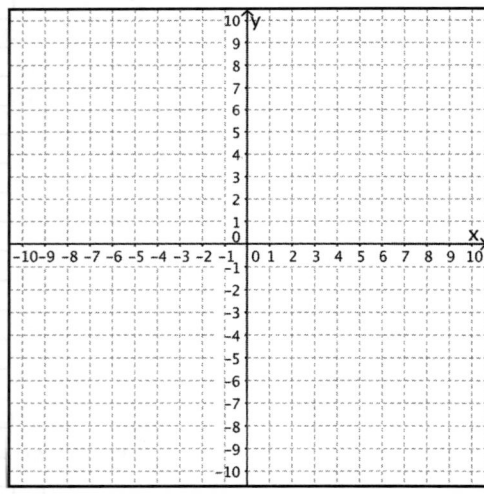

A (−2, −6)

B (−5, 2)

C (1, 4)

D (2, −2)

9. Which linear inequality is shown on the graph below?

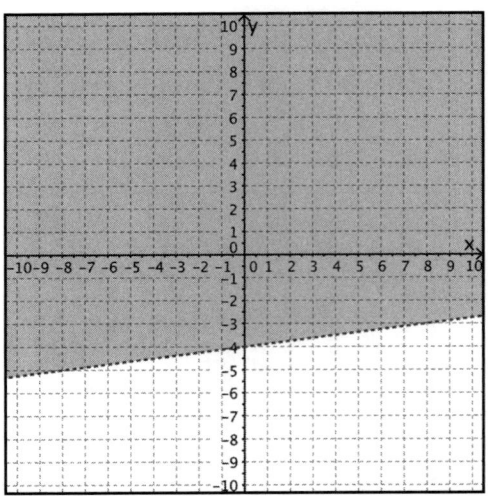

F $x + 8y < -32$

G $x + 8y \leq -32$

H $8x + y \geq -32$

J $x + 8y > -32$

10. Graph the inequality $7x + 3.5y \geq 28$

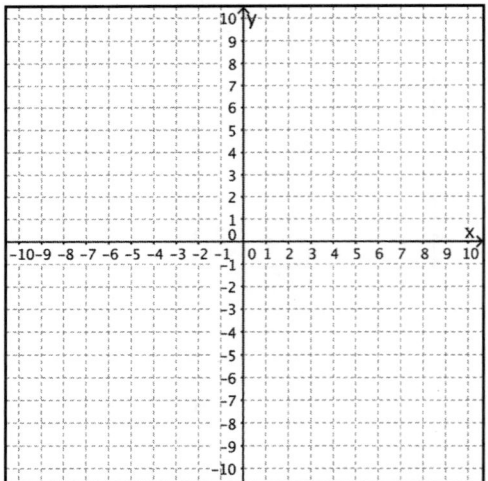

TRANSFORMATIONS OF LINEAR FUNCTIONS

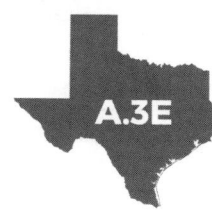

A.3E

The student is expected to determine the effects on the graph of the parent function $f(x) = x$ when $f(x)$ is replaced by $af(x)$, $f(x) + d$, $f(x - c)$, $f(bx)$ for specific values of a, b, c, and d.

ⓘ TELL ME MORE...

The linear parent function, $f(x) = x$, can be transformed by multiplying or adding real numbers to the independent variable, x. Different transformations of the parent function affect the graph of the parent function in different ways.

$$f(x) = x$$

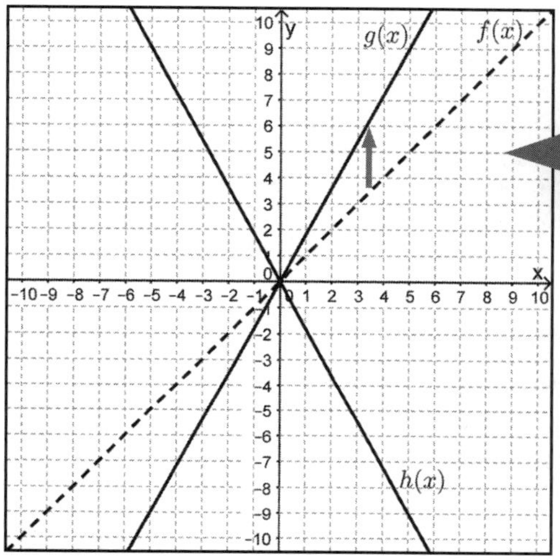

g(x) = af(x)
Vertical dilation by a factor of a. Reflection across the x-axis if $a < 0$.

g(x) = f(bx)
Horizontal dilation by a factor of $\frac{1}{b}$. Reflection across the y-axis if $b < 0$.

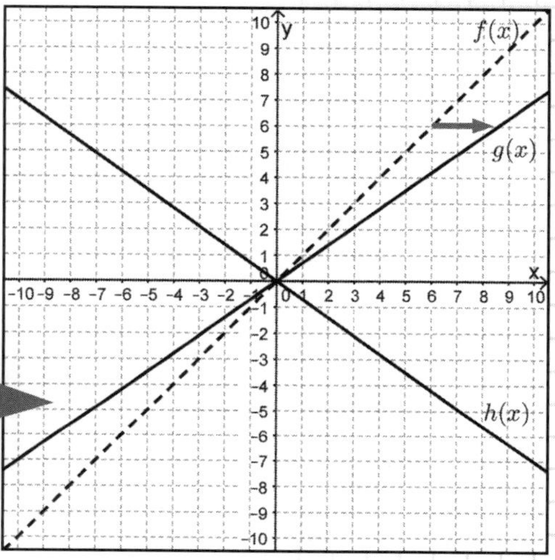

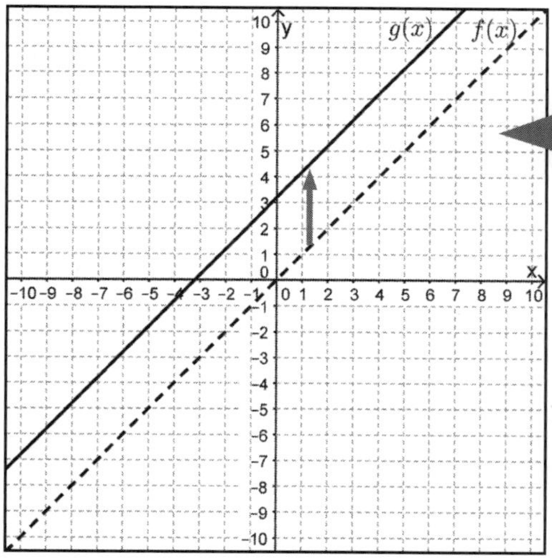

g(x) = f(x) + d
Vertical translation d units up if $d > 0$ and down if $d < 0$.

g(x) = f(x − c)
Horizontal translation c units to the right if $c > 0$ and to the left if $c < 0$.

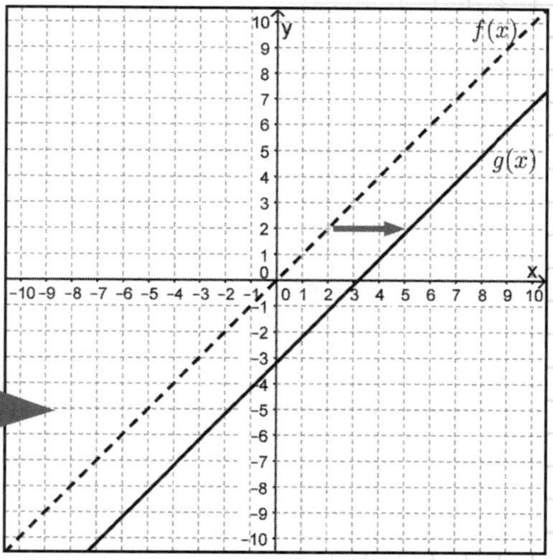

✓ EXAMPLES

EXAMPLE 1: Elena graphed $f(x) = x$ and $g(x) = f(x) - 7$ on her graphing calculator. How are the graphs of g and f related?

STEP 1 Identify the parameter that is added or multiplied to $f(x)$ to create $g(x)$. $g(x) = f(x) - 7$, so to create $g(x)$, you subtract 7 (or add −7) to $f(x)$. This addend is the parameter d.

$d = -7$

STEP 2 Determine the effect of the parameter d on the graph of $f(x)$.

The parameter d causes a vertical translation of d units. Since $d = -7$, there will be a vertical translation of $f(x)$ 7 units down to create the graph of $g(x)$.

Translate $f(x)$ 7 units down

STEP 3 Graph both f and g on the same screen or coordinate plane to confirm your answer.

The graph of f is translated, or shifted, 7 units down to create the graph of g.

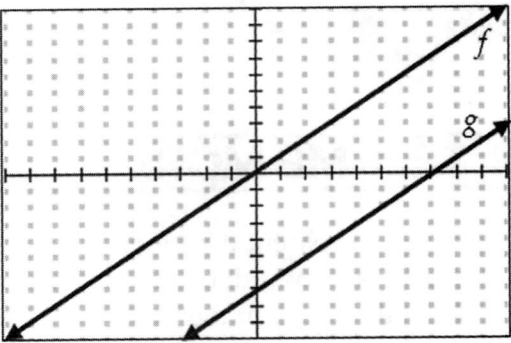

EXAMPLE 2: Daniel and his science lab group collected some data in a recent experiment. Daniel knows that he can use $k(x)$ to model the data if he takes the parent function, $f(x) = x$, changes the slope to 1.75 and the y-intercept to (0, 3.5). How does the slope and y-intercept of $k(x)$ compare to those of $f(x)$?

STEP 1 Identify the slope and y-intercept of the parent function, $f(x) = x$.

- $f(x) = x$ is equivalent to $f(x) = 1x + 0$
- $m = 1$ and $b = 0$

The slope of $f(x)$ is 1 and the y-intercept of $f(x)$ is (0, 0).

STEP 2 Determine how the slope changes from $f(x)$ to $k(x)$. The slope of $f(x)$ is 1 and the slope of $k(x)$ is 1.75. 1.75 is greater than 1, so the slope increases.

The graph of $k(x)$ will be steeper than the graph of $f(x)$

STEP 3 Determine how the y-intercept changes from $f(x)$ to $k(x)$. The y-intercept of $f(x)$ is (0, 0) and the y-intercept of $k(x)$ is (0, 3.5). The y-intercept for $f(x)$ moves upward 3.5 units to become the y-intercept of $k(x)$.

The y-intercept of $k(x)$ will be translated up from the y-intercept of $f(x)$.

YOU TRY IT!

Linear parent function $f(x) = x$ is graphed on a coordinate plane along with $h(x) = f(x - 6)$. How does the graph of h compare with the graph of f?

Graph:

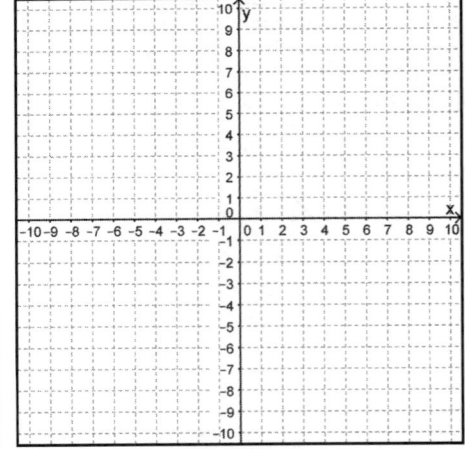

Does $h(x)$ include the parameter a, b, c, or d?

What transformation occurred to $f(x)$ to generate $h(x)$?

EXAMPLE 3 The graph of $f(x) = x$ and $g(x) = f(bx)$ are shown. What factor, b, is multiplied by x in order to generate the graph of g? What transformation does b create for g from f?

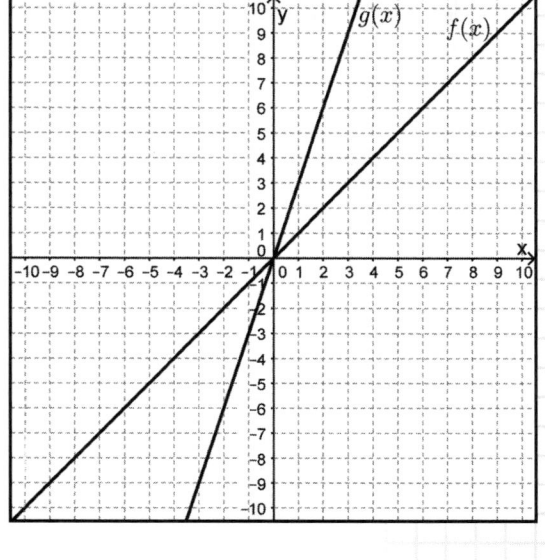

STEP 1 The parameter b, causes a horizontal dilation. Identify one point on each graph, f and g, with the same function value (y-coordinate).

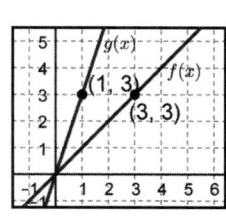

(3, 3) and (1, 3)

STEP 2 Determine the factor required to multiply by the x-coordinate of the point from $f(x)$ to get the x-coordinate of the point from $g(x)$. The parameter b is the reciprocal of this factor.

$f(x)$: (3, 3), so x-coordinate is 3

$g(x)$: (1, 3), so x-coordinate is 1

Multiply by $\frac{1}{3}$

b = 3

STEP 3 Determine the transformation.

The parameter b generates a horizontal dilation.

$b = 3$, so the dilation will be a factor of $\frac{1}{b} = \frac{1}{3}$, which is less than 1.

The dilation will be a compression.

The parameter b creates a horizontal compression of f by a factor of $\frac{1}{3}$.

PRACTICE

1. The graphs of lines m and n are plotted on the same coordinate plane. Line m represents the linear parent function. Line n represents a horizontal translation of line m 5 units to the right and a vertical stretch by a factor of 3. What are the equations for lines m and n?

2. The linear parent function, $f(x)$, is plotted on a graph along with $k(x)$. If $k(x) = f(2x) + 7$, how is the graph of $k(x)$ changed from $f(x)$?

3. The graph of $f(p)$ is shown. How will $f(p)$ change after a vertical translation of −4 units and a horizontal compression by a factor of $\frac{1}{2}$?

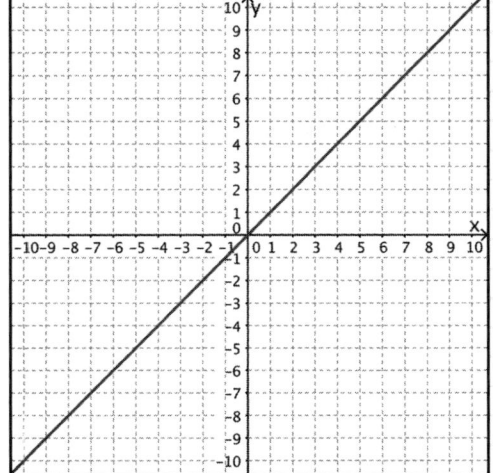

4. Function A, the linear parent function, and Function B are shown plotted on the graph. How was Function A transformed to create Function B?

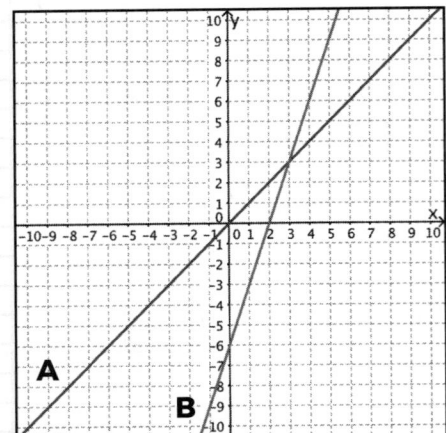

5. The table below represents points on the graph of a linear function that represents a transformation of the linear parent function $f(x)$.

x	$h(x)$
-5	-14
-1	-6
3	2
6	8

What transformations were applied to $f(x)$ to created $h(x)$?

6. The graph of $f(x) = x$ describes the linear parent function. If a horizontal translation of −6 units and a vertical compression by a factor of $\frac{1}{4}$ is applied to the graph of $f(x)$, how will the graph of $f(x)$ change?

7. Linear functions $h(x) = \frac{3}{2}x - 6$ and $g(x) = \frac{3}{4}(x + 1)$ are both graphed on a coordinate plane. How was the linear parent function, $f(x)$, transformed to created $h(x)$ and $g(x)$?

8. The linear parent function, $f(x)$, is plotted on a coordinate plane. A new function $h(x)$ represents the following transformation to $f(x)$.
$$h(x) = \frac{4}{5}f(x - 4)$$

How will the graph of $h(x)$ compare to the graph of $f(x)$?

9. Linear function $f(x) = x$ is plotted on a coordinate plane. The graph of $k(x)$ represents a transformation of $f(x)$ such that $k(x) = 4f(x - 3)$. Which of the following explains how the graph of $f(x)$ is changed to $k(x)$ with the transformations?

A The slope is steeper and the line shifts 3 units up.

B The slope is less steep and the line shifts 3 units left.

C The slope is steeper and the line shifts 3 units right.

D The slope is less steep and the line shifts 3 units down.

10. The graph of line p is represented using $f(x) = x$. The graph of line q is a transformation of line p with a vertical compression by a factor of $\frac{1}{2}$ and a vertical translation 8 units in the negative direction. Which of the following describes how line q compares to line p?

F Line q has a steeper slope and the y-intercept 8 units below line p.

G Line q has a slope that is less steep and the y-intercept 8 units above line p.

H Line q has a steeper slope and the y-intercept 8 units above line p.

J Line q has a slope that is less steep and the y-intercept 8 units below line p.

GRAPHING SYSTEMS OF LINEAR EQUATIONS

A.3F
A.3G

The student is expected to graph systems of two linear equations in two variables on the coordinate plane and determine the solutions if they exist.

The student is expected to estimate graphically the solutions to systems of two linear equations with two variables in real-world problems.

ℹ TELL ME MORE...

A **system of equations** is a set of two or more equations that describe a particular situation. If you represent an equation graphically, then you are representing all possible combinations of the variables x and y that make the equation true. For example, the graph of $2x - y = 4$ is shown. Every point on this line represents a pair of x- and y-values that make the equation true.

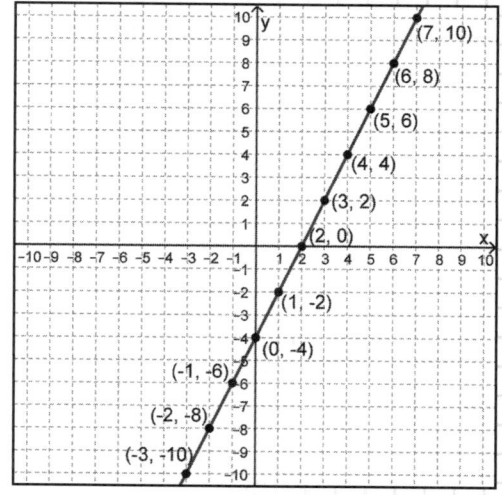

Ordered Pair (x, y)	Process	Result
(-3, -10)	$2(-3) - (-10) = -6 + 10$	4
(-2, -8)	$2(-2) - (-8) = -4 + 8$	4
(-1, -6)	$2(-1) - (-6) = -2 + 6$	4
(0, -4)	$2(0) - (-4) = 0 + 4$	4

If you were to graph two intersecting lines, then the same idea is true. Each line contains (x, y) values that make each equation true. The **intersection point** is a special point that makes both equations true. Hence, you can say that the intersection point is the **solution** to the system of equations. Consider the system shown. The only point that makes both equations true is (3, 2), which is the intersection point of their graphs.

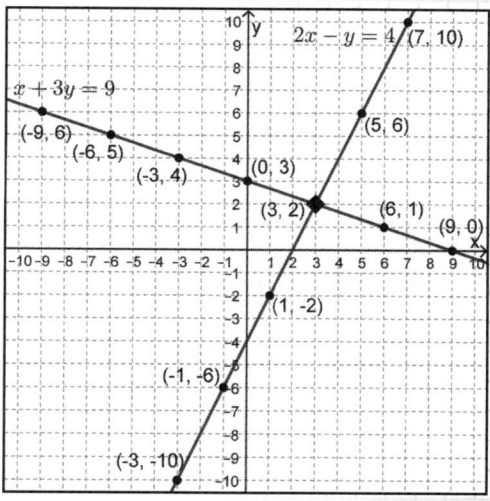

$$\begin{cases} 2x - y = 4 \\ x + 3y = 9 \end{cases}$$

Ordered Pair (x, y)	$2x - y = 4$	$x + 3y = 9$
(1, -2)	$2(1) - (-2) = 4$ True	$1 + 3(-2) = -5$ False
(0, 3)	$2(0) - 3 = -3$ False	$0 + 3(3) = 9$ True
(3, 2)	$2(3) - 2 = 4$ True	$3 + 3(2) = 9$ True
(6, 1)	$2(6) - 1 = 11$ False	$6 + 3(1) = 9$ True

So you can say that the solution to the system $2x - y = 4$ and $x + 3y = 9$ is (3, 2), or $x = 3$ and $y = 2$.

EXAMPLES

EXAMPLE 1: Graph the system $\begin{cases} y = 3x - 5 \\ x + 2y = 4 \end{cases}$. What is the solution to the system?

STEP 1 Graph both lines on the same coordinate plane. If you are using technology, graph one line as $Y1$ and the other line as $Y2$.

- $y = 3x - 5$ is in slope-intercept form. Plot the y-intercept $(0, -5)$ and then use the slope of 3 to plot the point $(0 + 1, -5 + 3) = (1, -2)$. Connect the points to make a line.

- $x + 2y = 4$ is in standard form. Determine the intercepts.

  - $x + 2(0) = 4$, so $x = 4$. The x-intercept is $(4, 0)$.

  - $(0) + 2y = 4$, so $2y = 4$ and $y = 2$. The y-intercept is $(0, 2)$.

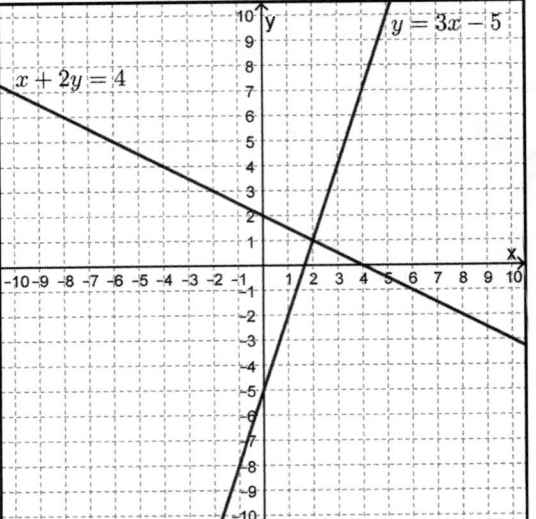

STEP 2 Locate the intersection point.
- The lines appear to cross at $(2, 1)$.

(2, 1)

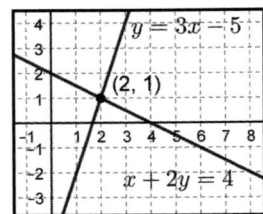

STEP 3 Verify algebraically that $(2, 1)$ makes both equations in the system true.

(2, 1) is the solution to the system.

$y = 3x - 5$	$x + 2y = 4$
$1 = 3(2) - 5$	$2 + 2(1) = 4$
$1 = 6 - 5$	$2 + 2 = 4$
$1 = 1$ ✔	$4 = 4$ ✔

EXAMPLE 2: Graph the system $\begin{cases} y = -\frac{2}{3}x + 4 \\ 2x + 3y = -3 \end{cases}$. What is the solution to the system?

STEP 1 Graph both lines on the same coordinate plane. If you are using technology, graph one line as $Y1$ and the other line as $Y2$.

- $y = -\frac{2}{3}x + 4$ is in slope-intercept form. Plot the y-intercept $(0, 4)$ and then use the slope of $-\frac{2}{3}$ to plot the point $(0 + 3, 4 - 2) = (3, 2)$. Connect the points to make a line.

- $2x + 3y = -3$ is in standard form. Determine the intercepts.

  - $2x + 3(0) = -3$, so $2x = -3$ and $x = -1.5$. The x-intercept is $(-1.5, 0)$.

  - $2(0) + 3y = -3$, so $3y = -3$ and $y = -1$. The y-intercept is $(0, -1)$.

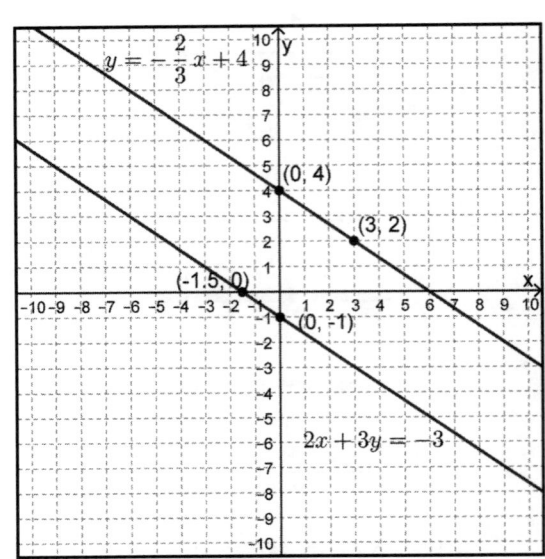

STEP 2 Locate the intersection point.

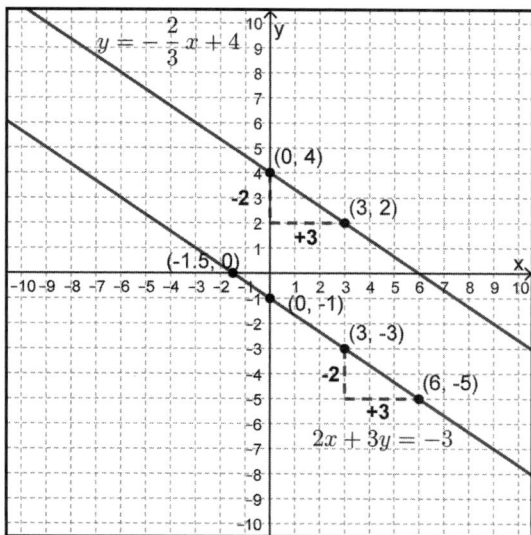

- The lines appear to be parallel. Verify graphically that the slope of both lines is $-\frac{2}{3}$. If two lines have the same slope, then they are indeed parallel.

- If two lines are parallel, then they will never intersect. If there is no intersection point, then there is no solution to the system.

There is no solution to the system.

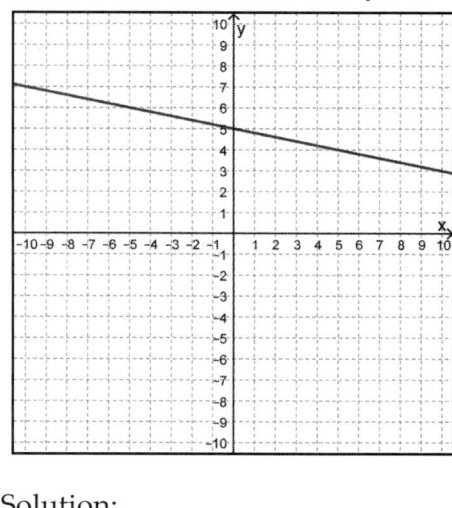

EXAMPLE 3 Townie Gym has two different membership plans. For one plan, there is a $75 membership fee and costs $15 per month. The second plan has no membership fee and costs $25 per month. The graph shows the costs of each membership plan as a function of time. Estimate the time when the two plans cost the same amount.

STEP 1 Estimate the location of the intersection point.

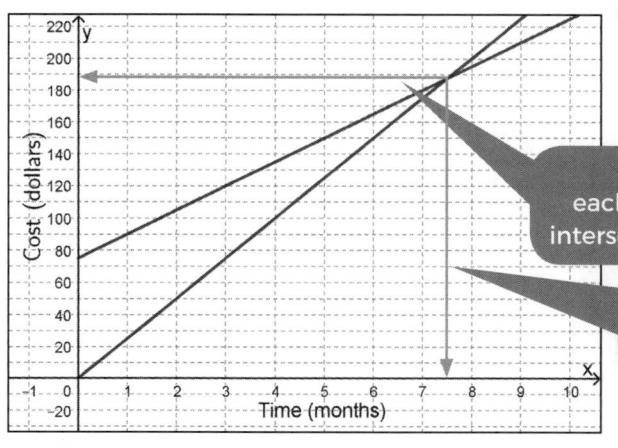

The y-axis is labeled by 20 and each gridline represents 10 dollars. The intersection point is between 180 and 190.

The x-axis is labeled by 1 and each gridline represents 1 month. The intersection point is between 7 and 8.

About (7.5, 185)

STEP 2 Interpret the intersection point as the estimated solution to the problem.

- The independent variable is time in months. So if $x \approx 7.5$, then the time is about 7.5 months.

- The dependent variable is cost in dollars. So if $y \approx 185$, then the cost is about $185.

At about 7.5 months, the cost will be about $185 for both membership plans.

PRACTICE

1. Graph the system of linear equations:

$$2x - 3y = 15$$
$$x + 2y = 4$$

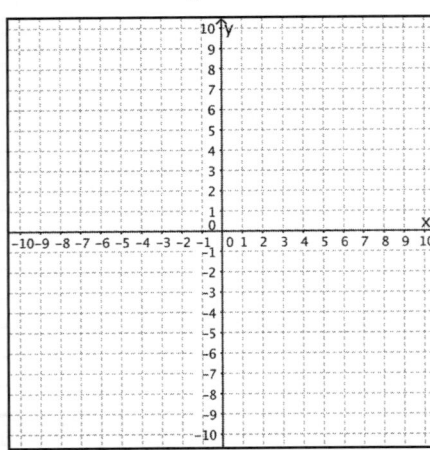

What is the solution to the system based on the graph?

2. The first equation in a system of linear equations is shown graphed below.

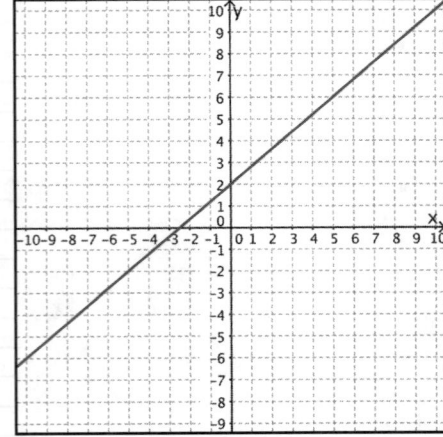

If the second equation has a slope of $-\frac{1}{2}$ and passes through the point $(-1, 9)$, what is the solution to the system of linear equations?

3. When the solution to the system of linear equations shown below is graphed, how many solutions will the system have?

$$3x + 2y = -4$$
$$2x + y = 1$$

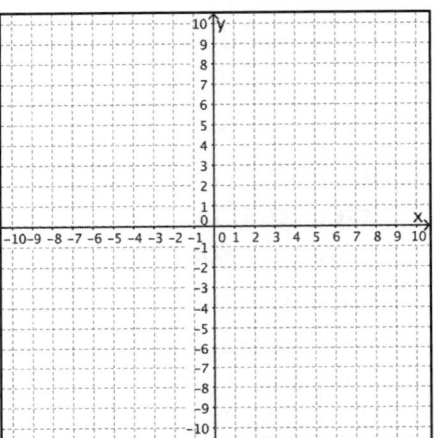

4. The graph below represents the first of two equations in a system of linear equations.

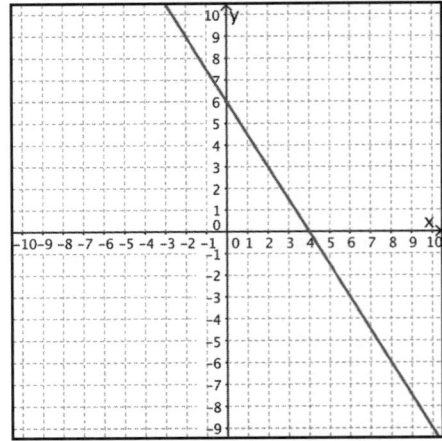

If the graph of the second equation in the system passes through the points $(3, -1)$ and $(-6, 5)$, identify the solution(s) if it exists.

5. A certain system of linear equations has no solution. One equation in the system is $3x - 4y = 24$. The second equation has a y-intercept of $(0, 5)$. What is the second equation in the system, written in standard form?

6. Candy is sold by the quarter pound at Pop's Candy Shoppe. One type of candy is $3.50 per quarter pound. A second type of candy costs $5.00 per quarter pound. Mariel bought a total of 2 pounds of the two candies to share with her best friend, Linda, and spent $31.00. If a system of linear equation is graphed representing the situation, what does the intersection point of the two lines on the graph represent?

7. Harriet graphed a system of linear equations and recorded some of the points for both lines in the following table.

x	Line A	Line B
-2	13	13
-1	11	11
0	9	9
1	7	7
2	5	5

What is the solution to the system of linear equations?

8. The line shown on the graph below is one equation in a system of linear equations.

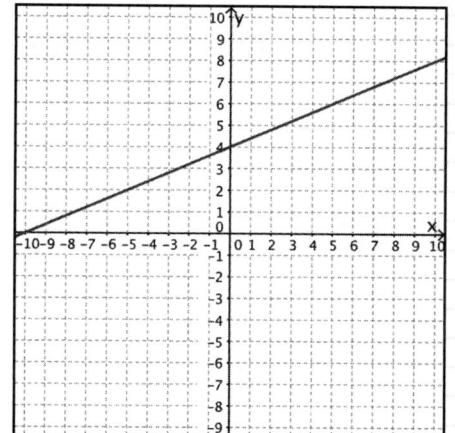

If the system of linear equations yields infinite solutions, which equation could be the second equation in the system?

A $y = -\dfrac{5}{2}x + 4$

B $y = \dfrac{5}{2}x + 4$

C $y = -\dfrac{2}{5}x + 4$

D $y = \dfrac{2}{5}x + 4$

9. Which of the following systems of linear equations would show two lines with one point of intersection when graphed?

F $6x - 8y = 48$
$-6x + 8y = 16$

G $y = -2x + 1$
$x = \dfrac{1}{2}y + \dfrac{1}{2}$

H $y = -x + 3$
$y = x - 3$

J $y - 5 = \dfrac{3}{4}(x - 4)$
$y - 2 = \dfrac{3}{4}(x - 3)$

10. Which graph can be used to find the solution to the system of equations below?

$$4x - 5y = 20$$
$$y = -\frac{1}{5}x + 1$$

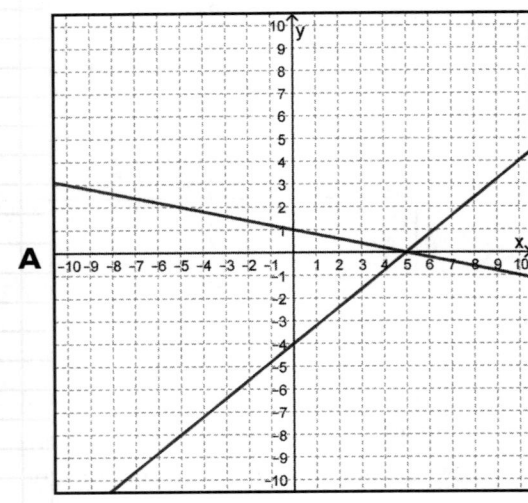

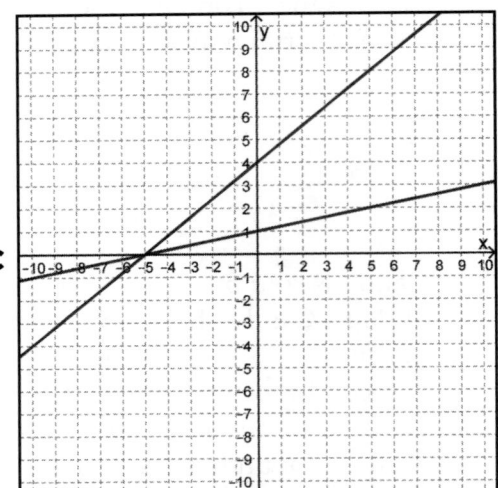

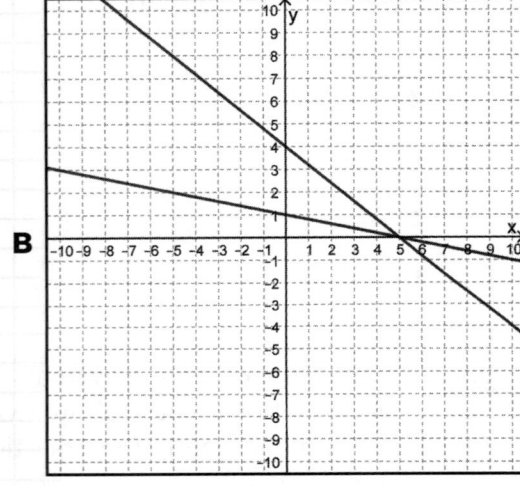

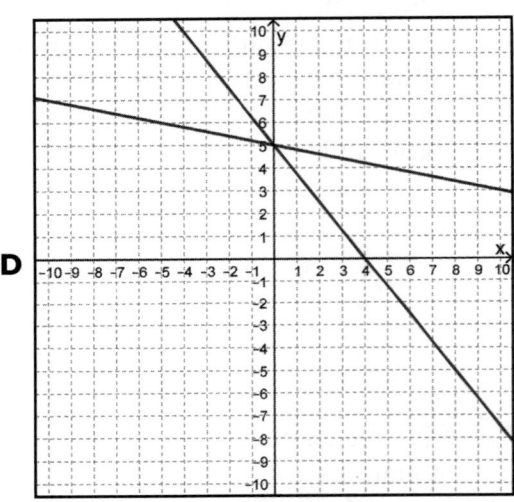

GRAPHING SYSTEMS OF INEQUALITIES

A.3H The student is expected to graph the solution set of systems of two linear inequalities in two variables on the coordinate plane.

ℹ TELL ME MORE...

The graph shows the system of two linear inequalities.

$$2x - 3y < 12$$
$$y \geq -2x + 4$$

The coordinate plane is divided into four regions.

- Region A is the region that is in the solution set for $2x - 3y < 12$ but not the solution set for $y \geq -2x + 4$. Point J lies in this region.

- Region B is the region that is in the solution set for $2x - 3y < 12$ and is in the solution set for $y \geq -2x + 4$. Since Region B is the overlap between the solution sets for both inequalities, it is the solution set for the system. Points C, D, E, and G lie in this region.

- Region C is the region that is not in the solution set for either inequality. Points A and I lie in this region. (Point A is on the boundary line for $2x - 3y < 12$. But since the inequality does not include "or equal to," the points along its boundary line are not included in its solution set.)

- Region D is the region that is in the solution set for $y \geq -2x + 4$ but not in the solution set for $2x - 3y < 12$. Points B and H lie in this region.

The boundary line for both inequalities, and the resulting intersection point, is important for determining the full solution set of the system of inequalities. If an inequality does not include "or equal to," then its boundary line is not a part of the solution set for the inequality. Hence, the boundary line (and its associated intersection point) cannot be included in the solution set for the system.

- A solid boundary line indicates that the points on the line are included in the solution set.

- A broken or dashed boundary line indicates that the points on the line are not included in the solution set.

✓ EXAMPLES

EXAMPLE 1: Graph the system of inequalities.

$$2x + 5y \geq 10$$
$$y < \frac{2}{3}x - 4$$

STEP 1 Since the first inequality is in standard form, determine the x-intercept and y-intercept of the boundary line.

(5, 0) and (0, 2)

For the x-intercept, $(x, 0)$, let $y = 0$.

$$2x + 5y = 10$$
$$2x + 5(0) = 10$$
$$2x = 10$$
$$x = 5$$

For the y-intercept, $(0, y)$, let $x = 0$.

$$2x + 5y = 10$$
$$2(0) + 5y = 10$$
$$5y = 10$$
$$y = 2$$

STEP 2 Graph the first inequality.

■ Plot the x- and y-intercepts on a graph. Connect them with a boundary line. The boundary line will be solid if the inequality includes "or equal to"or dashed if the inequality does not include "or equal to."

■ The inequality is ≥, or "greater than or equal to," so the boundary line will be solid.

$$2x + 5y \geq 10$$

■ Determine which region of the inequality to shade. Test the point (0, 0).

$2(0) + 5(0) \geq 10$

$0 \geq 10$, which is not true.

■ Since (0, 0) is not a part of the solution set for the first inequality, shade the region that does not contain (0, 0)

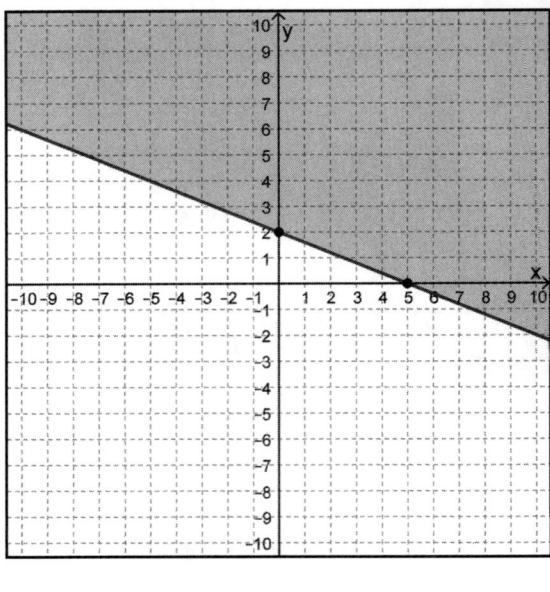

STEP 3 Graph the second inequality, $y < \frac{2}{3}x - 4$.

■ This inequality is in slope-intercept form.

■ Begin with the y-intercept, (0, −4).

■ Use the slope to plot a second point, (3, −2).

■ Determine which region of the inequality to shade. Test the point (0, 0).

$0 < \frac{2}{3}(0) - 4$

$0 < 0 - 4$

$0 < -4$, which is not true.

■ Since (0, 0) is not a part of the solution set for the second inequality, shade the region that does not contain (0, 0).

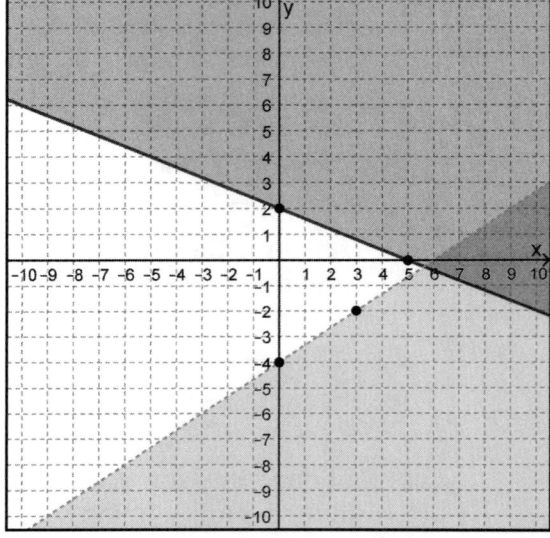

STEP 4 Determine the region that is the solution to the system of inequalities. The region where both solution sets overlap is the solution of the system.

Region B represents the solution of the system.

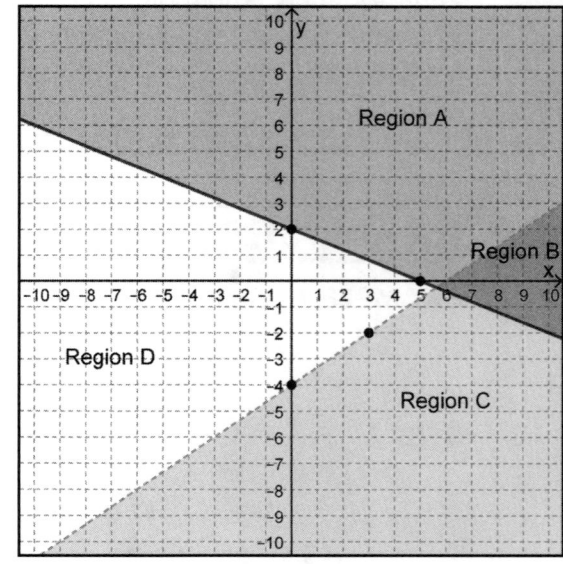

EXAMPLE 2: The graph of the following system of inequalities is shown.

$$y > \frac{1}{4}x - 5$$

$$y < 3x + 2$$

Which of the following points lie in the solution set of the system?

$A(4, -4)$	$D(10, 8)$
$B(0, 8)$	$E(-12, 2)$
$C(-16, -12)$	$F(2, -12)$

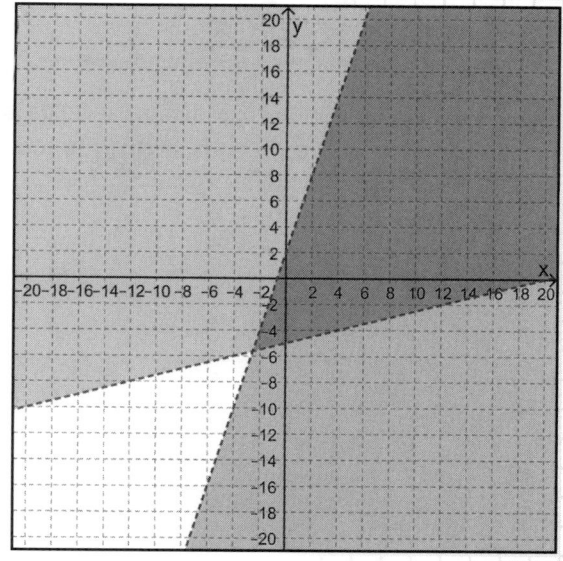

STEP 1 Plot the points on the coordinate plane.

STEP 2 Determine the region of the graph representing the solution to the system.

- Both inequalities do not include "or equal to," so the boundary lines are not a part of the solution set of the system.
- $y > \frac{1}{4}x - 5$, has a slope of $\frac{1}{4}$, so it is the line that is less steep.
- $y < 3x + 2$ has a slope of 3, so it is the line that is steeper.
- The point $(0, 0)$ makes both inequalities true, so the region containing $(0, 0)$ is the solution set for the system.

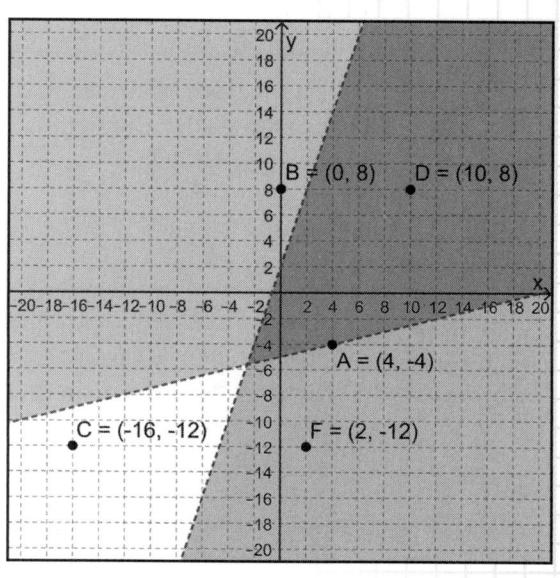

STEP 3 Identify the points that lie inside the solution set of the system.

- Point D lies in the region representing the solution set.
- Point A lies on the boundary line so it is not in the region representing the solution set.
- Point B is in the solution set for the inequality. $y < 3x + 2$ but not $y > \frac{1}{4}x - 5$, so it is not in the region representing the solution set

Only point D lies in the solution set of the system.

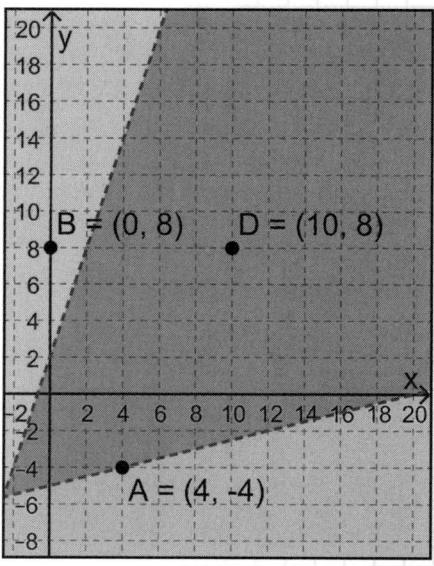

1. A system of linear inequalities is plotted on the graph below.

    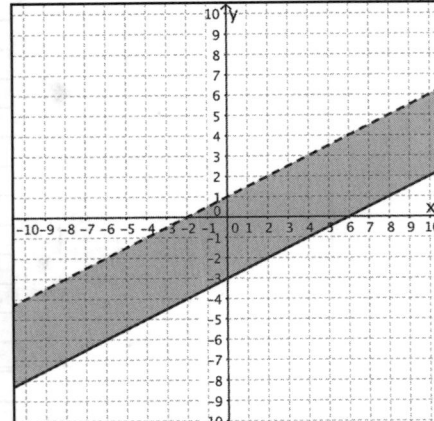

    What system of linear inequalities is represented by the graph.

2. The solution to a system of linear inequalities is shown on the graph. Which system does the graph best represent?

    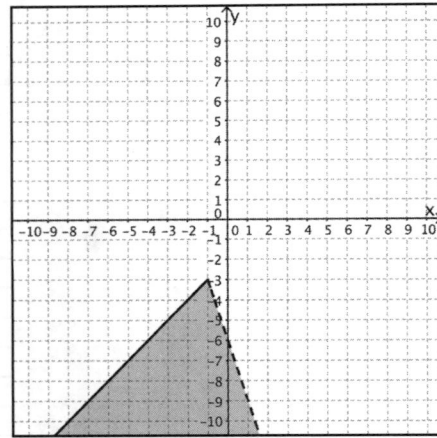

    F $y \leq x - 2$
       $y < -3x - 6$

    G $y \geq x - 2$
       $y < -3x - 6$

    H $y \leq x - 2$
       $y > -3x - 6$

    J $y < x - 2$
       $y \leq -3x - 6$

3. Use the graph to show which of the following points lies in the solution set for the system of linear inequalities shown.

    $$x + 2y < 4$$
    $$2x - y \leq 8$$

    $A\ (-2, -4)$ $B\ (6, 4)$ $C\ (-4, 4)$

    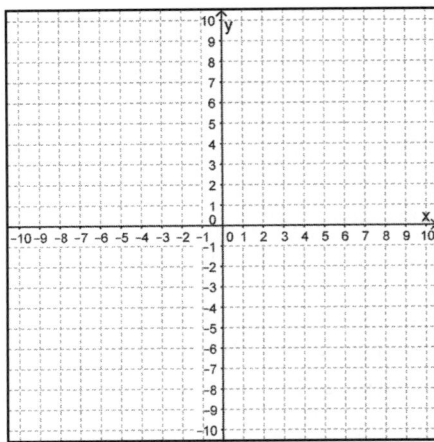

4. A system of linear inequalities is created by all points where the y-values are less than zero and the x-values are greater than zero. In which quadrant of the coordinate plane will the solution to the system of linear inequalities lie?

    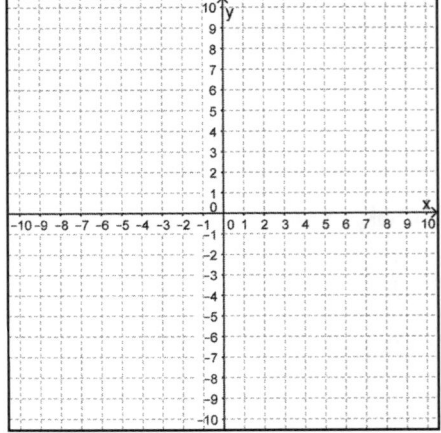

5. Susan has up to $60 to spend at the used book store annual sale and plans to buy at least 1 book. All hardback books cost $12 and all paperback books cost $4. This relationship is shown in the graph below.

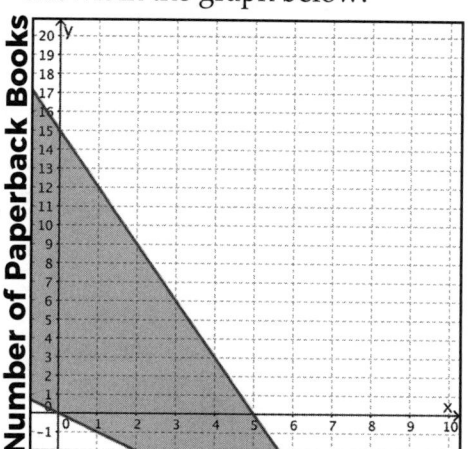

Number of Hardback Books

If Susan decides to buy 3 books in hardback form, what numbers of paperback books are possible for her to buy?

6. Rayna is starting a new exercise-training program. Each day she plans to either walk a 5-mile path or ride her stationary bike on a 10-mile program. Rayna's goal is to walk or ride more than 150 miles. This relationship is shown graphed below.

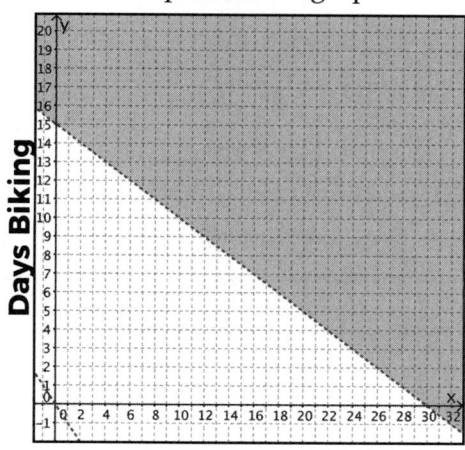

Days Walking

Rayna prefers walking so she plans to spend twice as many days walking as biking. What is the first point that falls within the solution set that satisfies these conditions?

7. A system of linear inequalities is shown partially graphed.

$$2x + y \le -5$$
$$3x - y < -2$$

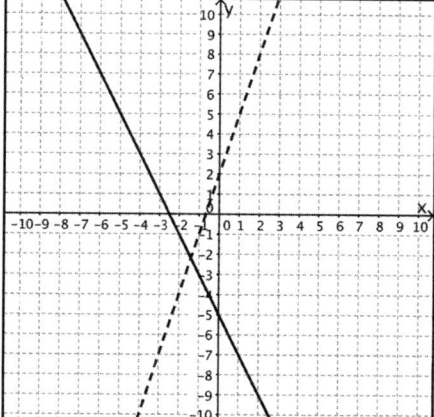

Which of the following ordered pairs lies in the solution set of the system of linear inequalities being graphed?

A $(-1, 3)$

B $(-2, -5)$

C $(3, -2)$

D $(-4, 3)$

8. On the graph below, shade to show the location of the solution to the system of linear inequalities.

$$4x - y \ge -3$$
$$x - y < 2$$

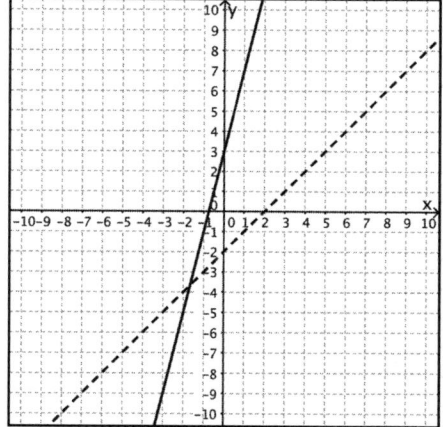

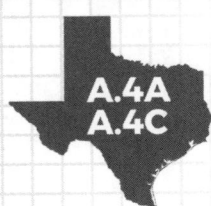

WRITING LINEAR FUNCTIONS FROM DATA

A.4A
A.4C

The student is expected to calculate, using technology, the correlation coefficient between two quantitative variables and interpret this quantity as a measure of the strength of the linear association.

The student is expected to write, with and without technology, linear functions that provide a reasonable fit to data to estimate solutions and make predictions for real-world problems.

ⓘ TELL ME MORE...

You can sometimes use a linear function to model real-world data. In those cases, you can then use the function to make predictions. There are several ways that you can generate a linear function model from data. Some of these methods may or may not use technology.

TREND LINE

If you have a table of data or a scatterplot of data that seems to follow a linear trend, then you can use a **trend line** to estimate the patterns in the data. In this process, you identify a starting point (y-intercept) and slope from the data to generate a linear function whose graph follows the trend in the data points reasonably well. You may or may not use technology to calculate the slope.

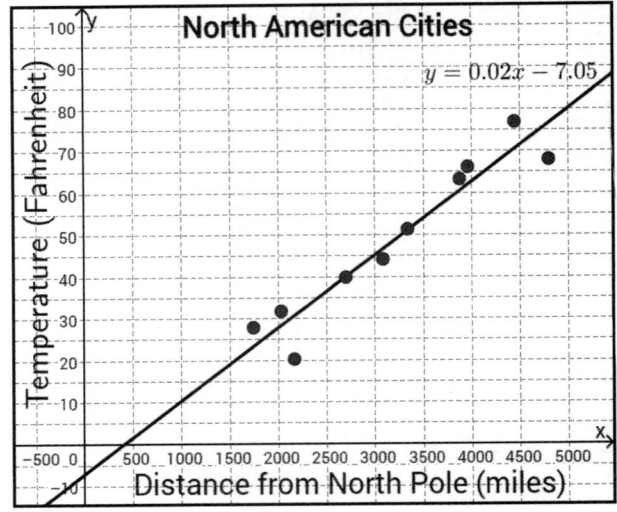

LINE OF BEST FIT

If a set of data seems to follow a linear trend, then you can use technology to calculate a **line of best fit** for the data. Statistical algorithms use a technique called **linear regression** to calculate the least distances between each data point and a linear model for the data. The **correlation coefficient**, r, generated by these algorithms tells you the strength of the linear fit for the data.

- If $r > 0$, the linear relationship has a positive correlation.
- If $r < 0$, the linear relationship has a negative correlation.
- If $|r| > 0.70$, then the relationship is a strong correlation.
- If $0.30 < |r| < 0.70$, then the relationship is a moderate correlation.
- If $|r| < 0.30$, then the relationship is a weak correlation ($r \approx 0$ is no correlation).

✓ EXAMPLES

EXAMPLE 1: The table shows the number of times that a cricket chirps in one second for a particular air temperature, in degrees Fahrenheit. Use technology to write a line of best fit for this data and interpret the correlation coefficient.

Number of Chirps, x	20	16	19.8	18.4	15	17.2	16	14.4
Temperature (°F), y	88.6	71.6	93.3	84.3	79.6	82.6	80.6	76.3

STEP 1 Select a graphing technology and enter the data into its data editor.

NORMAL FLOAT AUTO REAL RADIAN MP

L1	L2	L3	L4	L5	2
20	88.6	------	------	------	
16	71.6				
19.8	93.3				
18.4	84.3				
15	79.6				
17.2	82.6				
16	80.6				
14.4	76.3				
------	------				

L2(9)=

STEP 2 Use the technology's linear regression feature to calculate the line of best fit and correlation coefficient. Round as necessary.

$y = 2.732x + 35.397, r ≈ 0.851$

NORMAL FLOAT AUTO REAL RADIAN MP

LinReg
y=ax+b
a=2.731916244
b=35.39673223
r²=0.7234560755
r=0.8505622114

STEP 3 Use the correlation coefficient to determine the type and strength of the linear correlation.

$0.7 < 0.851 < 1$

Since r is between 0.7 and 1, there is a strong positive linear correlation.

EXAMPLE 2: The data in the graph show the year that different Native American pueblos were settled in the American Southwest and the year that the pueblos were abandoned. Use the scatterplot to write a trend line and estimate when a pueblos settled in 800 A.D. would have been abandoned.

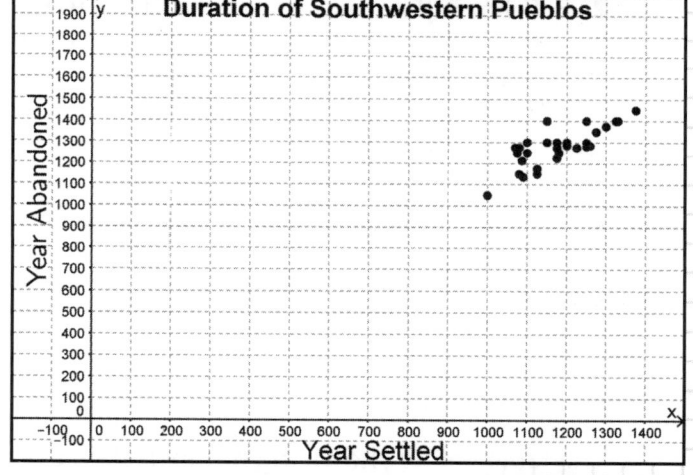

STEP 1 Identify the trend in the data and place a line that approximates this trend.

STEP 2 Determine two points near the center of the plot that you can use to calculate the slope of the trend line.

$$m = \frac{\Delta y}{\Delta x} = \frac{1400 - 1100}{1300 - 1000} = \frac{300}{300} = 1$$

$m = 1$

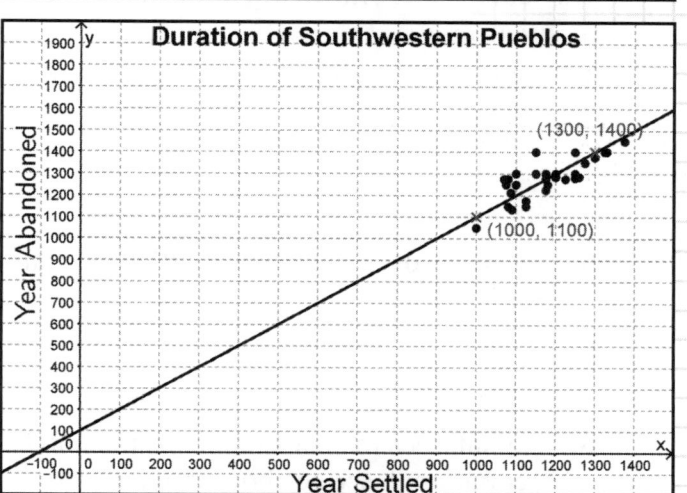

STEP 3   Use the slope of the trend line and either the y-intercept or one of the two points to determine the equation of the trend line.

y = x + 100.

$$y - y_1 = m(x - x_1)$$
$$y - 1100 = 1(x - 1000)$$
$$y - 1100 = x - 1000$$
$$y = x + 100$$

STEP 4   The year of settlement is the independent variable, x. Use the equation of the trend line to determine the function value for when $x = 800$.

$$y = x + 100$$
$$y = (800) + 100$$
$$y = 900$$

The settlement would likely have been abandoned in the year 900 A.D.

PRACTICE

Use the following information for Questions 1-3.

The scatterplot below shows the number of incorrect items on a quiz compared to the number of minutes spent studying for the quiz.

1.   What regression equation models a line of best fit for the data?

2.   What is the correlation coefficient, to the nearest hundredth, of the linear regression equation used to model the data?

3.   What does the correlation coefficient indicate about the data?

Use the following information for Questions 4-6.

The table below shows the number of weeks of practice for a competition and the score received.

4.   What regression equation models a line of best fit for the data?

5.   What is the correlation coefficient related to the fit of the trend line?

6.   What does the correlation coefficient for the regression equation describe about the data set?

Practice	Score
6	12.5
8	23
10	29.5
12	22
14	35
16	43.5
18	39.5

Use the following information for Questions 7-8.

Firefighters in one California county use data on flame length and fire speed to predict information about future grass fires. The graph below shows some of their collected data points.

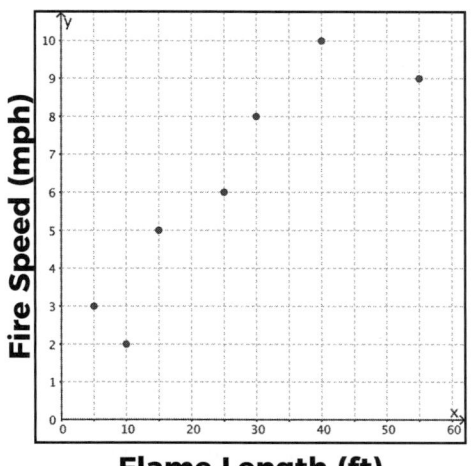

Flame Length (ft)

7. What linear function best models the data?

8. What length of flames can firefighters expect when the fire speed is 15 mph?

9. Which of the following correlation coefficients describes the strongest association between the variables?

   A −0.89

   B 0.74

   C 0.05

   D −0.57

Use the following information for Questions 10-11.

An investor is interested is in working with Flora's Flower Shop. Sales, in thousands of dollars, for the past 6 years are shown in the table.

Year, x	1	2	3	4	5	6
Sales, y	300	320	345	360	390	385

10. What linear function best models the relationship between sales in thousands of dollars, y, and the number of years, x, the business has been open?

11. If this trend continues, what will be the annual amount of sales for Flora's Flowers in year 10 of business?

12. Manny is a truck driver for a large company. The graph below shows Manny's distance from home given his time spent driving. What linear function best models the data for Manny's drive home?

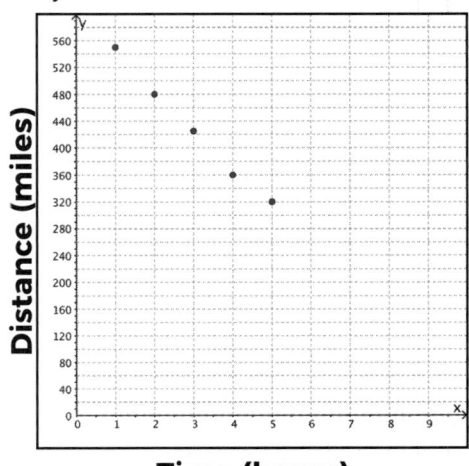

Time (hours)

   A $y = -87x + 637$

   B $y = -142\frac{1}{3}x + 692\frac{1}{3}$

   C $y = -58x + 600$

   D $y = -46x + 596$

ASSOCIATION AND CAUSATION

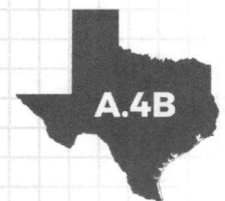

A.4B The student is expected to compare and contrast association and causation in real-world problems.

ⓘ TELL ME MORE...

Two or more measurable, quantifiable variables can be related in a number of ways. Sometimes, there is a **correlation** between two variables. Correlation means that as one variable increases or decreases, there is a pattern in how the other one changes. Sometimes, the correlation is **causation**, which means that changes in one variable actually and directly caused the changes in the second variable. If causation cannot be shown, then the correlation may be an **association**, which means that there is a relationship between the variables. However, in an association, it cannot clearly be shown that changes in the first variable cause changes in the second variable.

If you dig more deeply into the nature of that correlation, you can look at whether or not the change in one variable causes, or is merely related to, the change in the other variable.

ASSOCIATION

An **association** is a relationship or correlation between two variables. The change in one variable may or may not cause the change in the other.

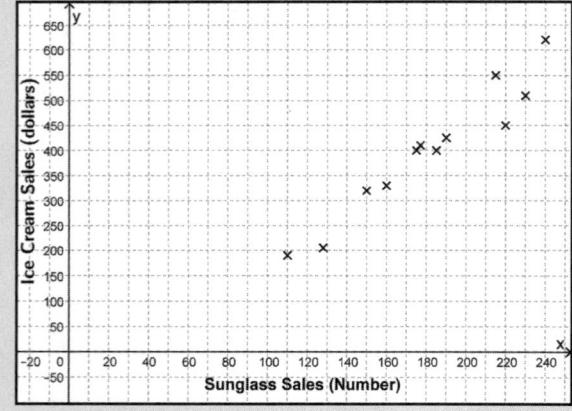

CAUSATION

A **causation** is a relationship between two variables in which a change in one variable actually causes the change (i.e., cause-and-effect relationship) in the second variable.

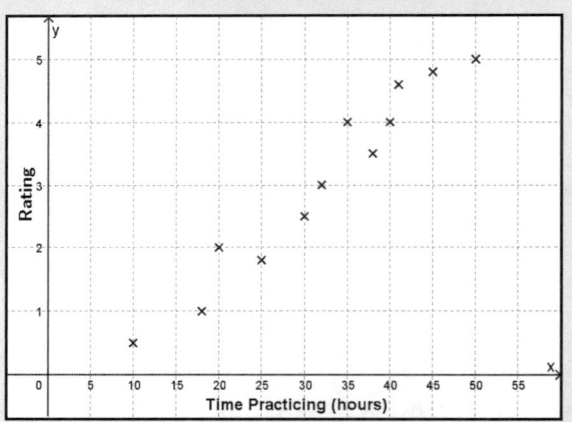

Sometimes, when two variables are related, there are **lurking variables** that also contribute to the change in the second variable. In the above example, ice cream sales may increase as the number of sunglasses sold increases; however, that does not mean that more people purchasing sunglasses causes ice cream sales to go up. Rather, the lurking variables of more sunshine and higher temperatures may be causing both variables to increase. However, studies can compare the amounts of time that different students who participate in band spend practicing and their ratings at a regional contest. The study design can account for other lurking variables to show that increasing the amount of time spent practicing will cause an increase in the student's rating at the regional contest.

EXAMPLE 1: Sandra collected data regarding the number of bus stops on a route and the amount of time it takes to run one complete route. Her data are shown.

Number of Bus Stops	10	12	12	15	15	17	18	20
Time for Complete Route (minutes)	20	25	30	40	35	40	45	50

Sandra concluded that as the number of bus stops increase, the amount of time for one complete route increases. Is this conclusion an association or a causation?

STEP 1 Write the conclusion in "if-then" form. Underline the "<u>if</u>" clause once and the "<u>then</u>" clause twice.

If the <u>number of bus stops increase</u>, then <u>the amount of time for one complete route increases</u>.

STEP 2 Identify any lurking variables that may exist.

- The amount of traffic may change throughout the day.
- The speed limit on different streets may change.
- The amount of time spent at each bus stop may change.

The amount of traffic, speed limit, and amount of time spent at each stop are possible lurking variables.

STEP 3 Consider whether or not the "if" variable is a cause and the "then" variable is an effect.

- Does an <u>increase in the number of bus stops</u> cause an <u>increase in the amount of time for one complete route</u>?
- If the bus stops more frequently, then it will spend more time during its route stopped and not moving.
- More time not moving, over the same length of the route, will add to the total time required to complete the route.

The relationship is a causation.

YOU TRY IT!

As the number of fire stations increases in a city, the response time for firefighters to arrive at a fire after they are called decreases. Is this relationship a causation? Explain.

If-then statement:

Possible lurking variables:

Causation (circle one)?   YES   NO

Explanation:

EXAMPLE 2: Ignacio noticed that when beach towel sales increased in his store that more people were sunburned. Is this relationship an association or a causation?

STEP 1   Write the conclusion in "if-then" form. Underline the "<u>if</u>" clause once and the "<u>then</u>" clause twice.

If <u>the number of beach towels sold increases</u>, then <u>the number of people with sunburns increases</u>.

STEP 2   Identify any lurking variables that may exist.

- More people go to the beach when the weather is warm and sunny.
- When the weather is warm and sunny, people are more likely to be outdoors.
- Spending more time in the sun increases the likelihood of getting a sunburn.

Warm temperatures, the amount of sunshine, and the amount of time spent in the sun are possible lurking variables.

STEP 3   Consider whether or not the "if" variable is a cause and the "then" variable is an effect.

- Does an <u>increase in the number of beach towels sold</u> cause an <u>increase in the number of people with sunburns</u>?
- If more people buy beach towels, that means there are more people going to the beach.
- More people going to the beach means that more people are outside.
- More people outside in the sun means that more people may get sunburned.
- Warm, sunny weather may lead to more people buying towels and to more people getting sunburned, so the lurking variable may be causing both beach towel sales and sunburn rates to increase.

The relationship is an association.

PRACTICE

For questions 1 – 3, determine if the statement implies association or causation.

1.   In a controlled experiment, researchers found a strong positive correlation coefficient in the data between the number of cigarettes smoked and the probability of developing lung cancer.

2.   Urban planners studied several large metropolitan areas and found that in cities with higher average summer temperatures the percent of homes with a pool was also higher.

3.   At Sofas and More, the manager created a scatterplot of average hours worked each week by each of the sales staff, x, and the average weekly sales in thousands of dollars, y. The scatterplot showed a 0.24 correlation coefficient between the variables.

4. Examine the graph below. Describe the relationship between the variables in terms of whether the graph shows association and/or causation.

Money Earned from Dog Walking

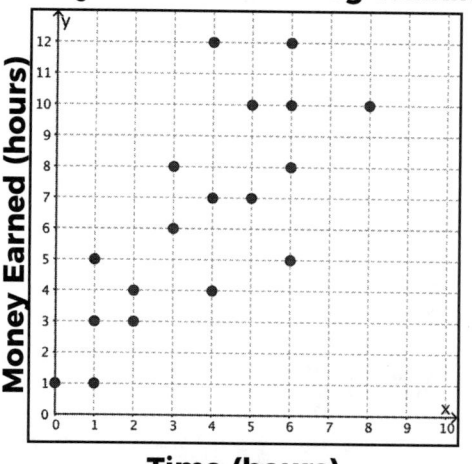

Time (hours)

5. A recent study used data to determine that the greater the number of hours spent at the gym, the less likely a person was to be a victim of a violent crime. Is this statement an association or causation? Explain.

6. In the data set relating test score, y, to the amount of time spent studying, x, the relationship has a correlation coefficient of 0.84. The data set is considered positively associated. Explain whether the data also indicate causation.

7. Which of the following statements would describe a situation that shows causation?

A The more often a person is tardy to work, the lower the person's salary will be.

B The longer the lights in the house are left on, the higher the electric bill will be.

C The lower the temperature outside, the more people found shopping at the mall.

D The more minutes spent exercising on the elliptical machine, the greater the number of calories burned.

8. Which of the following situations would NOT describe a causal relationship?

F The slower a person runs, the longer it will take the person to finish a race.

G The greater the distance a family travels on a car trip, the more time spent driving in the car.

H The faster a student types his research paper, the more pages his paper will be.

J The fewer questions answered correctly on a test, the lower the test score.

Unit 3
Writing and Solving Linear Functions, Equations, and Inequalities

DOMAIN AND RANGE OF LINEAR FUNCTIONS

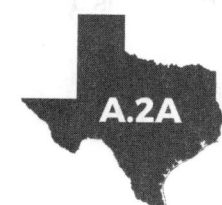

A.2A The student is expected to determine the domain and range of a linear function in mathematical problems; determine reasonable domain and range values for real-world situations, both continuous and discrete; and represent domain and range using inequalities.

ⓘ TELL ME MORE...

The **domain** of a function is the set of input values that are used for the independent variable. The **range** of a function is the set of output values for the dependent variable. For any linear function, $f(x) = ax + b$, the domain is the set of all real numbers and the range is the set of all real numbers.

However, when a linear function is used to model real-world situations the domain and range can be restricted to numbers that make sense for the situation. In these cases, the domain or range can be written in **inequality notation** for a **continuous** interval or as a **set of points** for a **discrete** situation.

A **continuous** subset of the domain or range of the full function includes a lower boundary point and an upper boundary point. For example, if you can only enter a bounce house if you are at least 36 inches but not more than 52 inches tall, then that interval can be represented with the inequality:

$$36 \leq x \leq 52$$

A **discrete** set of points in the domain or range involves a subset of only particular types of numbers, such as whole numbers or particular types of fractions. For example, if bags of cement mix weigh 25 pounds each, the weight of a collection of up to 5 cement bags can be represented with the set:

$$\{25, 50, 75, 100, 125\}$$

> **DOMAIN**
>
> For any linear function, $f(x) = ax + b$, the domain is the set of all real numbers.
>
> $$-\infty < x < \infty$$
>
> **RANGE**
>
> For any linear function, $f(x) = ax + b$, the range is the set of all real numbers.
>
> $$-\infty < f(x) < \infty$$

✓ EXAMPLES

EXAMPLE 1: A candy store at the mall has a shoppers rewards club. For joining the club, you receive 50 points. For every $20 you spend, you receive 4 points. If you spend x amounts of $20, then the function $b(x) = 4x + 50$ determines the number of points in your shoppers rewards account. Write the domain and range for this situation.

STEP 1 Determine if the situation is discrete or continuous.

- Points are awarded for increments of $20 spent, not for every dollar or fraction of a dollar.

- You cannot earn less than 1 whole point.

  Both input, x, and output, b(x), values are discrete.

STEP 2 Identify any upper or lower limits for each variable in the situation.

- The independent variable, the number of $20 increments that you spend (x), has a lower limit of 0 and no upper limit.
- The dependent variable, the number of points in your account ($b(x)$), has a lower limit of 50 and no upper limit.

x is at least 0 and $b(x)$ is at least 50.

STEP 3 State the domain of the function for this situation, keeping in mind the constraints already observed.

- x is discrete and has an increment of 1
- x is at least 0

The domain of the function in this situation is {0, 1, 2, 3, 4, 5, …} or the set of whole numbers.

STEP 4 State the range of the function for this situation, keeping in mind the constraints already observed.

- $b(x)$ is discrete and has an increment of 4
- $b(x)$ is at least 50

The range of the function in this situation is {50, 54, 58, 62, 66, 70, …}.

EXAMPLE 2: Plastic guards can be placed over hair clippers to control the length of the hair remaining. Each guard has a number, x, which is shown in the table along with its corresponding length remaining, $L(x)$. The table contains all of the plastic guards that come in a set with one pair of hair clippers. Write the domain and range for this situation.

Guard Number, x	Length Remaining, $L(x)$
1	$\frac{1}{8}$ in.
2	$\frac{1}{4}$ in.
3	$\frac{3}{8}$ in.
4	$\frac{1}{2}$ in.
5	$\frac{5}{8}$ in.
6	$\frac{3}{4}$ in.

STEP 1 Determine if the situation is discrete or continuous.

- Plastic guards are numbered with whole numbers.
- Lengths are for particular fractions, not every fraction.

Both input, x, and output, $L(x)$, values are discrete.

STEP 2 Identify any upper or lower limits for each variable in the situation.

- The independent variable, the guard number (x), has a lower limit of 1 and an upper limit of 6.
- The dependent variable, the length of hair remaining ($L(x)$), has a lower limit of $\frac{1}{8}$ inch and an upper limit of $\frac{3}{4}$ inch

x is between 1 and 6, inclusive, and $L(x)$ is between $\frac{1}{8}$ and $\frac{3}{4}$, inclusive.

STEP 3 State the domain of the function for this situation, keeping in mind the constraints already observed.

- x is discrete and has an increment of 1
- x begins at 1 and ends at 6

The domain of the function in this situation is {1, 2, 3, 4, 5, 6}.

STEP 4   State the range of the function for this situation, keeping in mind the constraints already observed.

- $L(x)$ is discrete and has an increment of $\frac{1}{8}$ inch.
- $L(x)$ begins at $\frac{1}{8}$ inch and ends at $\frac{3}{4}$ inch

The range of the function in this situation is $\left\{ \frac{1}{8}, \frac{1}{4}, \frac{3}{8}, \frac{1}{2}, \frac{5}{8}, \frac{3}{4} \right\}$

EXAMPLE 3: The graph of a part of linear function d is shown. Write the domain and range of the part shown using inequalities.

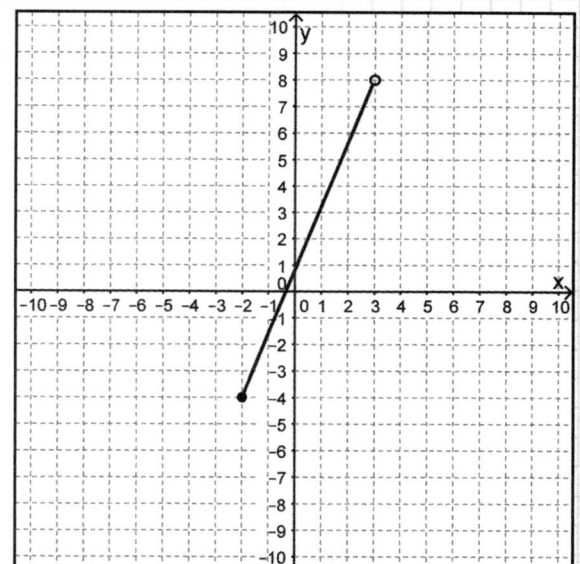

STEP 1   Determine the coordinates of each endpoint. There is a closed endpoint at (–2, –4) and an open endpoint at (3, 8).

(–2, –4) and (3, 8)

STEP 2   Identify the lower limit and upper limit for x using the x-coordinates of the endpoints. If the endpoint is an open endpoint (open circle), then that value is not included in the domain.

(–2, –4) and (3, 8)

> –2 is less than 3, so it is the lower limit. The endpoint is a closed circle so it is included in the domain.

> 3 is greater than –2, so it is the upper limit. The endpoint is an open circle so it is not included in the domain.

–2 is the lower limit (included) and 3 is the upper limit (not included)

STEP 3   Write the domain using the lower limit and the upper limit. If the limit is included in the domain, be sure the inequality contains "or equal to."

–2 ≤ x < 3

STEP 4   Identify the lower limit and upper limit for y using the y-coordinates of the endpoints. If the endpoint is an open endpoint (open circle), then that value is not included in the range.

(–2, –4) and (3, 8)

> –4 is less than 8, so it is the lower limit. The endpoint is a closed circle so it is included in the range.

> 8 is greater than –4, so it is the upper limit. The endpoint is an open circle so it is not included in the range.

–4 is the lower limit (included) and 8 is the upper limit (not included)

STEP 5   Write the range using the lower limit, inequalities, and the upper limit. If the limit is included in the range, be sure the inequality contains "or equal to."

–4 ≤ d(x) < 8

Write the domain and range for the situation using inequalities in Questions 1-4.

1. John's monthly natural gas bill is based on his gas usage at a rate of $0.804 per hundred cubic feet of gas, and a facilities charge of $7.75. The function $f(x) = 7.75 + 0.804x$, where x is the amount of gas consumed in hundreds of cubic feet, can be used to determine John's monthly gas bill.

2. Marlin's movie streaming service bill, $f(m)$, based on m number of movies watched can be found using the function $f(m) = 8 + 3.50m$. Marlin watches no more than 100 movies in a month.

3. A transportation app charges $2.50 plus $1.10 for each mile driven. The company delivers riders to any location within a 75-mile radius of the pick-up location. The function $f(m) = 2.50 + 1.10m$ can be used to determine the cost based on the number of miles driven.

4. The function $p(x) = 7500 + 150x$ represents the population of Hurricane Falls where x represents the number of years since 2000.

5. The graph below shows the cost of renting a compact car at a local rental agency based on the number of days rented.

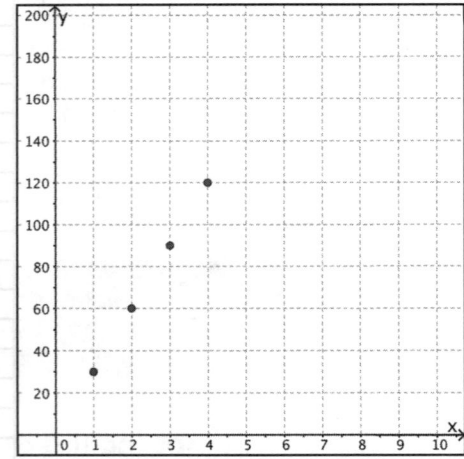

What are the possible range values for the situation?

6. A summer day camp is available to children ages 5 through 12. Campers pay a registration fee plus a daily rate for each day of camp attended. The camp meets 5 days a week for 9 weeks. The table below shows possible costs based on days of camp attended.

Days of Camp, x	Camp Cost ($), y
0	40
1	76
5	220
10	400
18	688

What are the domain and range values that are reasonable for the situation?

7. The function $g(x)$ is shown on the graph below.

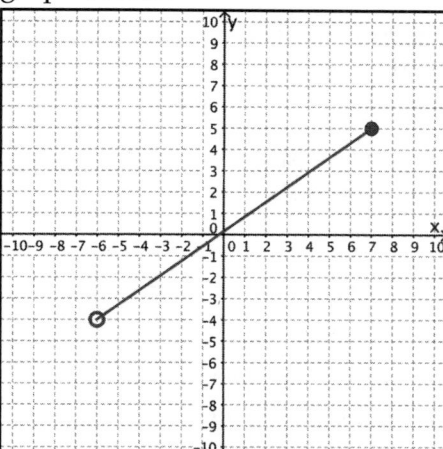

What is the domain of $g(x)$?

A $-6 < x \leq 7$

B $-6 \leq x < 7$

C $-4 < y \leq 5$

D $-4 \leq y < 5$

8. The local youth club is taking a group trip to a baseball game at a local stadium. They use a bus to transport the group members. The bus holds 35 people. The cost to attend the game is $18.50 per person and it will cost $15 to park the bus. The function $f(n) = 18.5n + 15$ can be used to describe the situation where n represents the number of people attending and $f(n)$ represents the total cost for the group trip. What values can be used in the domain of the function based on the situation?

F The set of real numbers greater than or equal to zero and less than or equal to 662.5

G The set of real numbers greater than or equal to 15 and less than or equal to 662.5

H The set of integers greater than or equal to 1 and less than or equal to 35

J The set of integers greater than 1 and less than or equal to 15

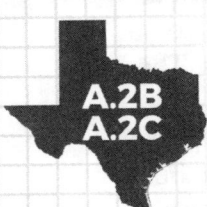

A.2B
A.2C

The student is expected to write linear equations in two variables in various forms, including $y = mx + b$, $Ax + By = C$, and $y - y_1 = m(x - x_1)$, given one point and the slope and given two points.

The student is expected to write linear equations in two variables given a table of values, a graph, and a verbal description.

(i) TELL ME MORE...

You can represent **linear functions** using graphs, tables, and verbal descriptions of the relationship. From these representations, you can identify either one point and the slope of the line or two points. With this information, you can then write the equation. There are several different forms of linear equations. Usually, any form will work but depending on the information that you are given, one form may be more efficient than the others.

Slope-Intercept Form	Standard Form	Point-Slope Form
$y = mx + b$	$Ax + By = C$	$y - y_1 = m(x - x_1)$
• m represents the slope of the line	• A, B, and C are typically integers	• m represents the slope of the line
• b represents the y-coordinate of the y-intercept, $(0, b)$	• Used when two quantities are combined	• The point (x_1, y_1) represents any point on the line
• Used when you know one point (especially the y-intercept) and the slope	• Useful when you know the x-intercept and y-intercept	• Used when you know one point and the slope
• Used when you know the starting point and rate of change in a word problem		• Used when you know two points and use them to calculate the slope

You can use a verbal description, table, or graph to determine the slope or the coordinates of a point. The graph, table, and sentence below all describe the same line. Use the information presented to write the slope and y-intercept in the space provided.

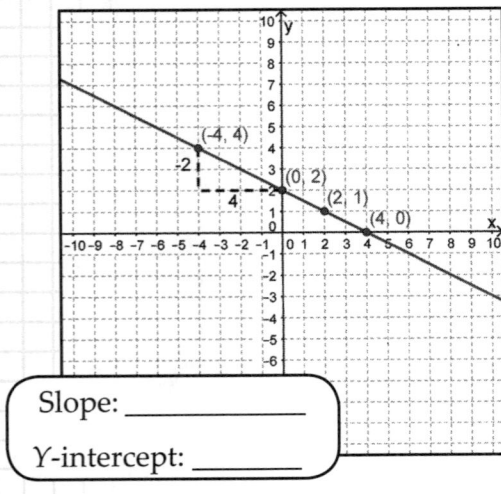

$0 - (-4) = 4$

x	y
-4	4
0	2
2	1
4	0

$2 - 4 = -2$

Slope: _____

Y-intercept: _____

Slope: _____

Y-intercept: _____

A line passes through the points (-4, 4) and (0, 2).

Slope: _____

Y-intercept: _____

 **EXAMPLES**

EXAMPLE 1: The table represents some points on the graph of a linear function.

x	-2	0	1	5
y	-8	-5	-3.5	2.5

Write a function that represents the same relationship.

STEP 1 A table contains sets of points for the function. Select 2 points (pairs of x- and y-values).

(0, –5) and (–2, –8)

STEP 2 Calculate the slope using the slope formula. Let $(-2, -8) = (x_1, y_1)$ and $(0, -5) = (x_2, y_2)$.

$m = \dfrac{3}{2}$

$m = \dfrac{y_2 - y_1}{x_2 - x_1} = \dfrac{-5 - (-8)}{0 - (-2)} = \dfrac{3}{2}$

STEP 3 One of the points in the table, $(0, -5)$, is the y-intercept. Substitute $m = \dfrac{3}{2}$ and $b = -5$ into slope-intercept form.

$y = \dfrac{3}{2}x - 5$

$y = mx + b$
$y = \dfrac{3}{2}x + (-5)$

EXAMPLE 2: The table represents some points on the graph of a linear function. Write an equation in standard form that represents the same relationship.

x	y
-2	15
-1	-12.5
2	-5
3	-2.5
6	5

STEP 1 A table contains sets of points for the function. Select 2 points (pairs of x- and y-values).

(2, –5) and (6, 5)

STEP 2 Calculate the slope using the slope formula. Let $(2, -5) = (x_1, y_1)$ and $(6, 5) = (x_2, y_2)$.

$m = \dfrac{5}{2}$

$m = \dfrac{y_2 - y_1}{x_2 - x_1} = \dfrac{5 - (-5)}{6 - 2} = \dfrac{10}{4} = \dfrac{5}{2}$

STEP 3 None of the points in the table represent an intercept. Instead, substitute $m = \dfrac{5}{2}$, $x_1 = 2$, and $y_1 = -5$ into the general point-slope form.

$y + 5 = \dfrac{5}{2}(x - 2)$

$y - y_1 = m(x - x_1)$
$y - (-5) = \dfrac{5}{2}(x - 2)$

STEP 4 Simplify the expression and then write it in standard form.

–5x + 2y = –20

$y + 5 = \dfrac{5}{2}(x - 2)$
$y + 5 = \dfrac{5}{2}x - 5$
$2y + 10 = 5x - 10$
$-5x + 2y = -20$

EXAMPLE 3: Write an equation in standard form that can be represented by the graph shown.

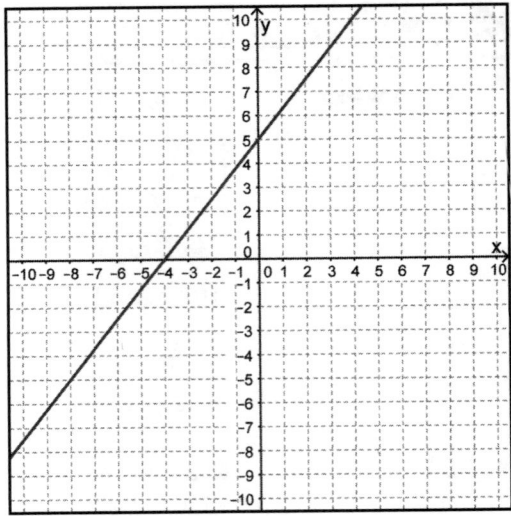

STEP 1 Identify two points on the graph. The *x*-intercept, (−4, 0), and *y*-intercept, (0, 5) can be used.

(−4, 0) and (0, 5)

STEP 2 Count the slope in the graph using Δ*x*, the change in *x* between the two points, and Δ*y*, the change in *y* between the two points.

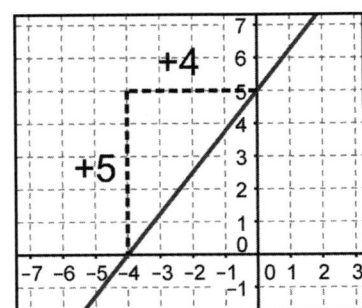

$$m = \Delta x - \Delta y = \frac{5}{4}$$

$$m = \frac{5}{4}$$

STEP 3 Substitute the slope for *m* and the *y*-coordinate of the *y*-intercept for *b* into slope-intercept form.

$$y = \frac{5}{4}x + 5$$

$$y = mx + b$$
$$y = \frac{5}{4}x + 5$$

STEP 4 Simplify the equation and write it in standard form.

−5x + 4y = 20

$$y = \frac{5}{4}x + 5$$
$$4y = 5x + 20$$
$$-5x + 4y = 20$$

PRACTICE

Answer the questions that follow.

1. The table below shows some points on the graph of a linear function. Write the function in slope-intercept form.

x	y
-4	2
0	8
2	11
3	12.5
6	17

2. The table below shows some points on the graph of a linear function. Write the function in standard form.

x	y
-2	6.5
0	5
4	2
6	0.5
10	-2.5

3. The table below shows some points on the graph of a linear function. Write the function in point-slope form.

x	-3	0	6	10
y	-6.5	-5	-2	0

4. Monty wants to plan a bowling party with his friends for his birthday. The table below shows the possible costs, y, based on the number of people, x, attending the party. What equation also describes the relationship between the cost and the number of party guests?

x	y
-4	2
0	8
2	11
3	12.5
6	17

5. Mr. Pearlman's phone bill is based on a flat fee plus a cost per 100 text messages sent each month. The table below shows possible amounts of the phone bill, y, related to the number of text messages, x, Mr. Pearlman sends during a billing period.

x	0	15	22	30
y	18	21	22.4	24

Which function also represents the same linear relationship?

A $y - 18 = \frac{1}{5}x$

B $y - 18 = 5x$

C $y = \frac{1}{5}(x - 18)$

D $y = 5(x - 18)$

For problems 6 – 8 use the following graph.

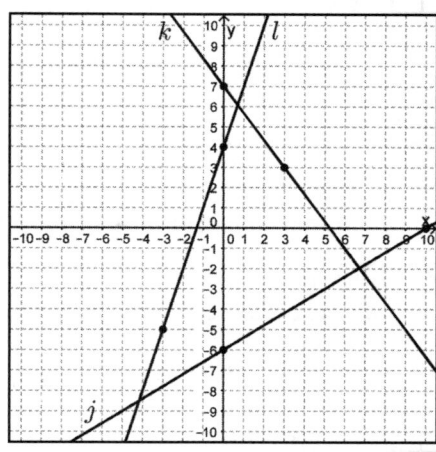

6. What equation written in standard form represents line j on the graph?

7. What equation written in slope-intercept form represents line k on the graph?

8. What equation written in point-slope form represents line l shown on the graph?

9. The graph below shows the growth of Ching's plant for her science experiment over a number of days. What linear equation in slope-intercept form represents the graph of the growth of Ching's plant over time?

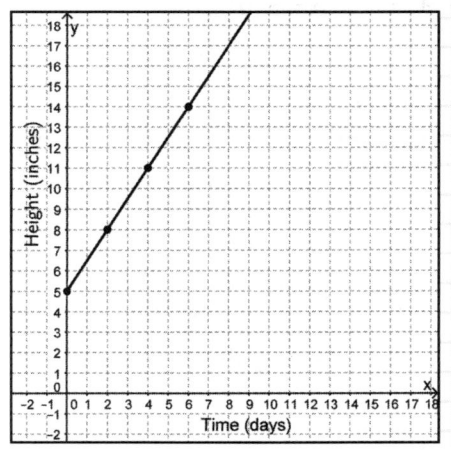

WRITING AND SOLVING LINEAR FUNCTIONS, EQUATIONS, AND INEQUALITIES **105**

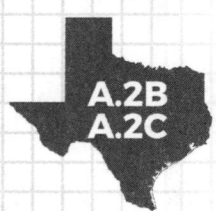

A.2B A.2C

The student is expected to write linear equations in two variables in various forms, including $y = mx + b$, $Ax + By = C$, and $y - y_1 = m(x - x_1)$, given one point and the slope and given two points.

The student is expected to write linear equations in two variables given a table of values, a graph, and a verbal description.

(i) TELL ME MORE...

You can represent **linear functions** using graphs, tables, and verbal descriptions of the relationship. From these representations, you can identify either one point and the slope of the line or two points. With this information, you can then write the equation. There are several different forms of linear equations. Usually, any form will work but depending on the information that you are given, one form may be more efficient than the others.

Slope-Intercept Form	Standard Form	Point-Slope Form
$y = mx + b$	$Ax + By = C$	$y - y_1 = m(x - x_1)$
• m represents the slope of the line	• A, B, and C are typically integers	• m represents the slope of the line
• b represents the y-coordinate of the y-intercept, $(0, b)$	• Used when two quantities are combined	• The point (x_1, y_1) represents any point on the line
• Used when you know one point (especially the y-intercept) and the slope	• Useful when you know the x-intercept and y-intercept	• Used when you know one point and the slope
• Used when you know the starting point and rate of change in a word problem		• Used when you know two points and use them to calculate the slope

A verbal description in a mathematical context will state one point and the slope, one point and information allowing you to infer the slope, or two points. From there, you have enough information to write the linear equation. A verbal description in a real-world context may give you information about quantities that are being combined or about a starting point with a rate of change.

✓ EXAMPLES

EXAMPLE 1: Write an equation in standard form that passes through the point (−4, 5) and has a slope of $\frac{7}{4}$.

STEP 1 Decide which formula to use. Since you are given the coordinates of one point and the slope, point-slope form is an appropriate choice.

Use point-slope form, $y - y_1 = m(x - x_1)$.

STEP 2 Substitute the slope for m, the x-coordinate of the given point for x_1, and the y-coordinate of the given point for y_1.

$$y - y_1 = m(x - x_1)$$
$$y - 5 = \frac{7}{4}(x - (-4))$$

$y - 5 = \frac{7}{4}(x + 4)$

STEP 3 Simplify the equation and write it in standard form.

$-7x + 4y = 48$

$$y - 5 = \frac{7}{4}(x - (-4))$$
$$y - 5 = \frac{7}{4}x + 7$$
$$4y - 20 = 7x + 28$$
$$-7x + 4y = 48$$

EXAMPLE 2: Write an equation in slope-intercept form that passes through the points $(7, -3)$ and $(1.5, 8)$.

STEP 1 Decide which formula to use. Since you are given the coordinates of two points, you can calculate the slope and use point-slope form.

Use point-slope form, $y - y_1 = m(x - x_1)$.

STEP 2 Calculate the slope using the slope formula. Let $(7, -3) = (x_1, y_1)$ and $(1.5, 8) = (x_2, y_2)$.

$$m = \frac{y_2 - y_1}{x_2 - x_1} = \frac{8 - (-3)}{1.5 - 7} = \frac{11}{-5.5} = 2$$

$m = -2$

STEP 3 Substitute the slope for m, the x-coordinate of the given point for x_1, and the y-coordinate of the given point for y_1.

$$y - y_1 = m(x - x_1)$$
$$y - (-3) = -2(x - 7)$$

$y + 3 = -2(x - 7)$

STEP 4 Simplify the equation and write it in slope-intercept form.

$$y + 3 = -2(x - 7)$$
$$y + 3 = -2x + 14$$
$$y = -2x + 14 - 3$$

$y = -2x + 11$

> **YOU TRY IT!**
>
> Write the equation of a line in standard form that passes through the points $(4, 9)$ and $(0, -7)$.
>
> Which form (slope-intercept, point-slope, or standard form) should you use to begin with? Why?
>
> _____
>
> _____
>
> _____
>
> What is the slope of the line? _____
>
> Substitute what you know, simplify, and write the equation in standard form.
>
> _____

EXAMPLE 3: At Barker's Burger Stand, hamburgers cost $4.50 each and one order of French fries costs $1.75. Sales tax is included in the prices. Let h represent the number of hamburgers purchased and f represent the number of orders of French fries purchased. Mrs. Okanabe paid $34 for some hamburgers and French fries for her family. Write an equation in standard form that describes what Mrs. Okanabe purchased.

STEP 1 Determine the information that is given in the problem. Highlight the information if desired.

- Hamburgers cost $4.50 and h represents the number of hamburgers purchased.

- French fries cost $1.75 and f represents the number of orders of French fries purchased.

Write an expression describing the cost of the hamburgers purchased. Write an expression describing the cost of the French fries purchased.

$$\text{Cost per Item} \times \text{Number of Items Purchased} = \text{Cost for Items}$$
$$\$4.50 \times h = 4.50h \text{ for the cost of the hamburgers}$$
$$\$1.75 \times f = 1.75f \text{ for the cost of the French fries}$$

4.50h and 1.75f

STEP 3 Combine these costs to write a linear equation in standard form.

$$\text{Cost of Hamburgers} + \text{Cost of French fries} = \text{Amount Paid}$$
$$4.50h + 1.75f = 34$$

4.50h + 1.75f = 34

PRACTICE

1. What equation written in standard form represents a linear function that passes through the point (1, 3) and has a slope of $-\frac{1}{2}$?

2. What linear equation written in point-slope form represents a line that passes through the points (−6, 1) and (−5, 4)?

3. What linear equation written in slope-intercept form represents a line with intercepts at (0, −4) and (2, 0)?

4. Joaquin graphs a line that passes through the point (5, −2) and has a slope of −3. Write an equation in standard form describing Joaquin's line.

5. Ms. Calhoun is an architect. On one of her sets of plans she draws two lines. The first line has a slope of $\frac{4}{3}$ and an x-intercept of (−6, 0). The second line is parallel to the first line and passes through the point (6, 5). What equation in point-slope form represents Ms. Calhoun's second line? Hint: parallel lines have the same slope.

6. On the graph of a certain line, $m = \frac{2}{5}$ and $b = -3$. What is the equation of the line in standard form

7. Reba and her best friend Kimi are joining a yoga studio. There is a membership fee of $25 to join the studio, plus the friends will each pay $2.50 per day for classes each time they attend. Write an equation in standard form that each friend can use to determine her total yoga expense based on the number of days of class attended.

9. Mrs. Kim took her son and his friends to the movies. Before and during the movie she made several trips to the concession stand to buy sodas and popcorn for the boys. The sodas she bought were priced at $4.50 each and the popcorn at $6.75 each. Mrs. Kim spent a total of $74.25. If s represents the number of sodas purchased and p represents the number of containers of popcorn purchased, which equation does NOT describe Mrs. Kim's purchases at the movies?

A $p = -\frac{2}{3}s + 11$

B $4.5s + 6.75p = 74.25$

C $p = -\frac{3}{2}s + 16.5$

D $p - 5 = -\frac{2}{3}(s - 9)$

8. The local senior center is planning to take a group to see a musical production. The performance tickets are regularly priced at $40 for adults, but anyone over 65 receives a discounted ticket price of $32. The senior center has raised a total of $800 in donations to afford the trip. What linear equation, written in both standard and slope-intercept forms, would model the number of regular adult tickets, x, and the number of senior tickets, y, that can be purchased for the trip?

10. Demontré spots a hot air balloon one morning on his way to work. The hot air balloon is 650 feet over the ground when Demontré spots it and is descending to a landing area at a rate of 22 feet per minute. Let x represent the number of minutes since Demontré saw the balloon and y represents the height of the balloon above the ground. Which equation models the height of the hot air balloon given the number of minutes since Demontré first saw it?

F $y = 22x + 650$

G $y = -22x + 650$

H $y = -22x - 650$

J $y = 22x - 650$

APPLYING DIRECT VARIATION

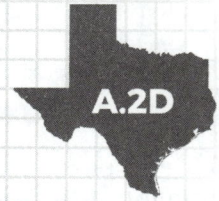

A.2D The student is expected to write and solve equations involving direct variation.

ⓘ TELL ME MORE...

Linear functions are relationships with a constant rate of change, often written in slope-intercept form, $y = mx + b$. Proportional relationships are a special class of linear functions where there is a constant rate of change but the y-intercept is $(0, 0)$. Proportional relationships can be written in general form, $y = kx$, where k represents a constant of proportionality. The constant of proportionality is equivalent to the slope of a linear function.

The function $y = kx$ is a collection of ordered pairs (x, kx). The table and graph show how the function values, y, are all multiples of the constant of proportionality, k.

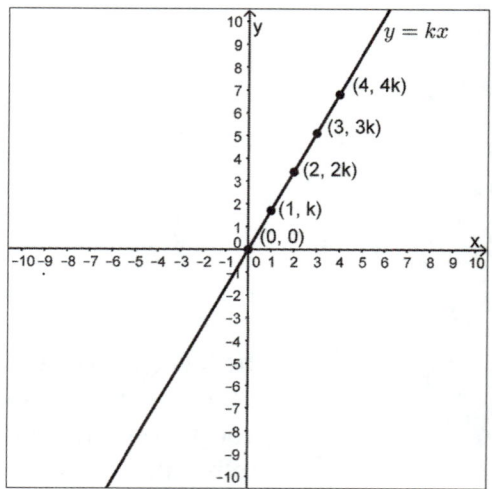

x	0	1	2	3	4
$y = kx$	0	$k(1) = 1k$	$k(2) = 2k$	$k(3) = 3k$	$k(4) = 4k$

Direct variation is a special way of writing proportional relationships where the dependent variable *varies directly* as the independent variable. "Directly" means that the independent variable is multiplied by a **constant of variation**, k, which is equivalent to the constant of proportionality or slope of the line generated by y. As such, the ratio of y to x is constant.

$$y = kx \longrightarrow \frac{y}{x} = k$$

Because direct variation is related to proportional relationships, you may sometimes see direct variation relationships written as "directly proportional." So if x represents the independent variable and y represents the dependent variable, the following two statements have the same meaning.

y varies directly as x

y is directly proportional to x

✓ EXAMPLES

EXAMPLE 1: The value of y is directly proportional to x. If $x = 25$ when $y = 110$, what is the value of y when $x = 40$?

STEP 1 The <u>dependent variable</u> is directly proportional to the <u>independent variable</u>. Use the given x-y pair of values to write the equation, $y = kx$, with an equation having the variable, k.

$y = kx$
$110 = k(25)$
$110 = 25k$

110 = 25k

STEP 2 Solve the equation for k.

$k = 4.4$

STEP 3 Use the constant of variation, k, to write $y = kx$.

$y = 4.4x$

$$110 = 25k$$
$$110 \div 25 = 25k \div 25$$

STEP 4 Substitute $x = 40$ into the direct variation equation and simplify.

$y = 176$

$$y = 4.4k$$
$$y = 4.4(40)$$

EXAMPLE 2: The mass of a substance varies directly as its volume. The mass of 48 grams of cork has a volume of 200 cubic centimeters. How many grams of cork has a volume of 150 cubic centimeters? Record your answer and fill in the bubbles on your answer document.

STEP 1 The <u>dependent</u> variable is directly proportional to the <u>independent variable</u>. Use the given values of mass and volume to write the equation, $m = kV$.

$$m = kV$$
$$48 = k(200)$$
$$48 = 200k$$

$48 = 200k$

STEP 2 Solve the equation for k.

$$48 = 200k$$
$$48 \div 200 = 200k \div 200$$

$k = 0.24$

STEP 3 Use the constant of variation, k, to write $m = kV$.

$y = 0.24x$

STEP 4 Substitute $V = 150$ into the direct variation equation and simplify.

$$m = 0.24V$$
$$m = 0.24(150)$$

$y = 36$

STEP 5 Since the question is a gridded response question, enter your response on the grid provided. Practice using the grid with the instructions.

1. Record a 3 in the first column containing numbers. Record a 6 in the next column.

2. Bubble the **3** beneath the numeral 3. Bubble **6** beneath the numeral 6.

YOU TRY IT!

The value of y varies directly as x. If $y = 75$ when $x = 10$, what is the value of x when $y = 60$?

Value of k: _____

Equation: _____

Value of x for $y = 60$: _____

EXAMPLE 3: The value of y varies directly with the value of x. Write a function that represents the relationship between x and y if $y = \frac{25}{6}$ when $x = 10$.

STEP 1 The <u>dependent variable</u> is directly proportional to the <u>independent variable</u>. $\quad y = kx$
Use the given values of x and y to write the equation, $y = kx$.
$$\frac{25}{6} = 10k$$

$y = kx$
$\frac{25}{6} = k(10)$
$\frac{25}{6} = 10k$

STEP 2 Solve the equation for k.
$$k = \frac{5}{12}$$

$\frac{25}{6} = 10k$
$\frac{25}{6} \div 10 = 10k \div 10$
$\frac{25}{6} \times \frac{1}{10} = k$

STEP 3 Use the constant of variation, k, to write $y = kx$.
$$y = \frac{5}{12}x$$

PRACTICE

Answer the questions that follow.

1. Izzy was examining photo she had made of herself. The measurements of Izzy in real life are directly proportional to her measurements in the photo. In the photo Izzy is 3.5 inches tall and 1.2 inches wide at the shoulders. In real life Izzy is 25.2 inches wide at the shoulders. How tall is Izzy in real life?

2. Carley is using a map to estimate the length in miles of her upcoming road trip. On her map $\frac{1}{8}$-inch represents 10 miles. What length on a map represents the distance of 248 miles Carley plans to travel to her first destination?

3. The weekly salary, s, Joaquin earns at his job varies directly with the number of hours worked, h. What equation can be used to determine Joaquin's weekly salary if his rate of pay is $7.50 per hour?

4. The cost of a call to India, c, varies directly with the number of minutes of talk time, m. What equation can be used to find the cost of a call that is m minutes long if the phone company charges $3.52 per minute of talk time? If a call cost $105.60, how many minutes were spent talking?

5. The weight of an object on Earth is directly proportional to the object's weight on the moon. The graph below shows some weights for objects on both the Earth and the moon.

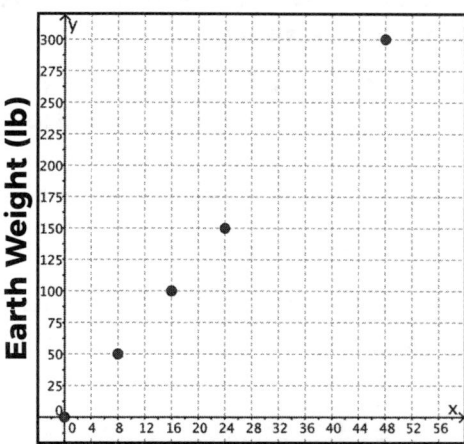

Moon Weight (lb)

Based on this information, what is the weight of an object on Earth that weighs 65 pounds on the moon?

6. In the table below, the values of y are directly proportional to the values of x.

x	0	1	3	4
y	0	3	9	12

What is k, the constant of proportionality between the values in the table?

7. In a direct variation relationship, $y = 14$ when $x = 4$. What is k, the constant of variation in the situation?

8. In a relationship, the value of y is directly proportional to the value of x. If $y = 18$ when $x = 8$, what is y when $x = 20$?

9. The relationship between the mass of an object in grams and the object's weight in pounds is shown in the table below.

Weight (pounds), x	Mass (grams), y
2	908
5	2,270
8	3,632
10	4,540
15	6,810

What function represents the direct proportional relationship between an object's mass in grams as related to its weight in pounds?

A $y = 454x$

B $y = \frac{1}{454}x$

C $y = x + 454$

D $y = 454^x$

10. Hookes's Law for an elastic spring states that the distance a spring stretches varies directly as the force applied. If a force of 16 pounds stretches a certain spring 9 inches, how much will a force of 40 pounds stretch the spring?

F $4\frac{4}{9}$ inches

G $22\frac{1}{2}$ inches

H 33 inches

J $71\frac{1}{9}$ inches

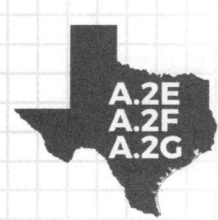

A.2E
A.2F
A.2G

The student is expected to write the equation of a line that contains a given point and is parallel to a given line.

The student is expected to write the equation of a line that contains a given point and is perpendicular to a given line.

The student is expected to write an equation of a line that is parallel or perpendicular to the *x*- or *y*-axis and determine whether the slope of the line is zero or undefined.

(i) TELL ME MORE...

The slope of a line tells you the steepness of the graph. If two lines are parallel, then they never intersect and have the same steepness. If two lines are perpendicular, then they intersect at a right angle. The slopes of parallel and perpendicular lines have special relationships

PARALLEL LINES

Parallel lines have the same slope. Because the slope, or steepness, of the lines is the same, they will never intersect. If *a*, *b*, and *c* are real numbers, then the lines y_1 and y_2 are parallel.

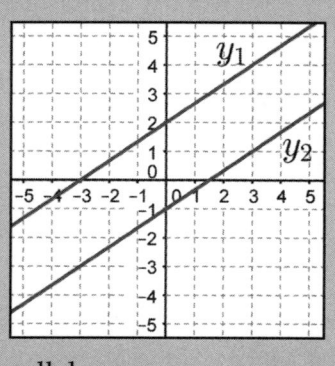

$$y_1 = ax + b$$
$$y_2 = ax + c$$

PERPENDICULAR LINES

Perpendicular lines intersect at a right angle. Because of the coordinate relationships, the slopes of perpendicular lines are negative reciprocals of each other. If *a*, *b*, and *c* are real numbers, then the lines y_3 and y_4 are perpendicular.

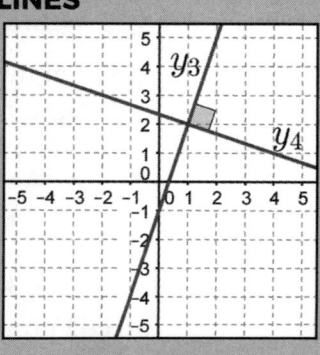

$$y_3 = ax + b$$
$$y_4 = -\frac{1}{a}x + c$$

✓ EXAMPLES

EXAMPLE 1: The graph of line *k* is shown. Write the equation, in slope-intercept form, of line *j* that is parallel to line *k* and has a *y*-intercept of (0, −3).

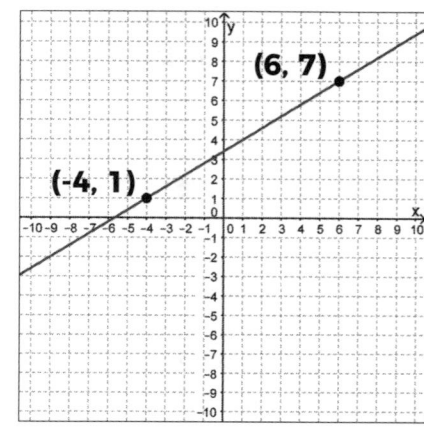

STEP 1 Determine the slope of line *m* using the slope formula and the two points given in the graph.

$$m = \frac{y_2 - y_1}{x_2 - x_1} = \frac{7 - 1}{6 - (-4)} = \frac{6}{10} = \frac{3}{5}$$

slope of line *k*: $m = \frac{3}{5}$

STEP 2 Since line *j* and line *k* are parallel, they have the same slopes.

slope of line *j*: $m = \frac{3}{5}$

STEP 3   Use the slope of line j along with the y-coordinate of the given y-intercept to write the equation of line j in slope-intercept form.

$$y = mx + b$$
$$y = \frac{3}{5}x + (-3)$$

$$\mathbf{y = \frac{3}{5}x - 3}$$

EXAMPLE 2: The graph of line f is shown. Write the equation, in standard form, of line h that is perpendicular to line f and passes through the point (1, 1).

STEP 1   Determine the slope of line f using the slope formula and the two points given in the graph.

$$m = \frac{y_2 - y_1}{x_2 - x_1} = \frac{4 - (-3)}{4 - 2} = \frac{7}{2}$$

slope of line f: $m = \frac{7}{2}$

Points shown on graph: **(4, 4)** and **(2, -3)**

STEP 2   Since line f and line h are perpendicular, they have slopes that are negative reciprocals of each other.

$$\text{Slope of line } h = -\frac{1}{\text{Slope of line } f} = -\frac{1}{\left(\frac{7}{2}\right)} = -\frac{2}{7}$$

slope of line h $m = -\frac{2}{7}$

STEP 3   Use the slope of line h along with the given point, (1, 1), to write the equation of line h using point-slope form.

$$y - y_1 = m(x - x_1)$$
$$y - 1 = -\frac{2}{7}(x - 1)$$

$$\mathbf{y - 1 = -\frac{2}{7}(x - 1)}$$

STEP 4   Simplify the equation and write it in standard form.

$$y - 1 = -\frac{2}{7}(x - 1)$$
$$y - 1 = -\frac{2}{7}x + \frac{2}{7}$$
$$7y - 7 = -2x + 2$$
$$2x + 7y - 7 = 2$$
$$2x + 7y = 9$$

$$\mathbf{2x + 7y = 9}$$

YOU TRY IT!

The table contains points that lie on the graph of a linear function.

x	-2	2	4	5
y	-8.5	-3.5	-1	0.25

What is the slope of the line represented?

_____

Line n is parallel to this line and passes through the point (2, 5).

What is the slope of line n?

_____

Write the equation of line n in slope-intercept form.

EXAMPLE 3: Thao is using computer software to design a quilt. She plots the point (6.5, 4) and wants to draw line c that passes through the point and is parallel to the x-axis and line d that passes through the point and is perpendicular to the x-axis. Write the equation and determine the slope of both line c and line d.

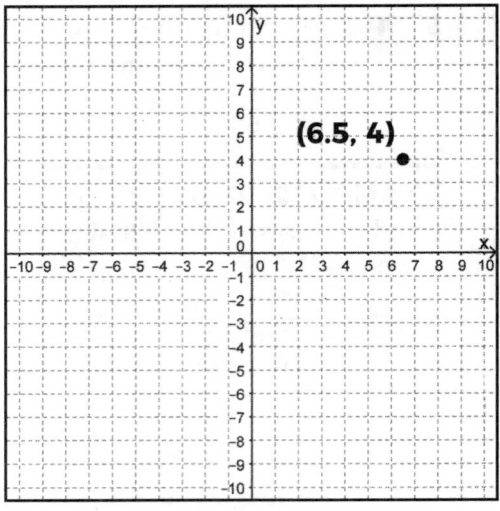

STEP 1 Draw line c through (6.5, 4) that is parallel to the x-axis. Draw line d through (6.5, 4) that is perpendicular to the x-axis.

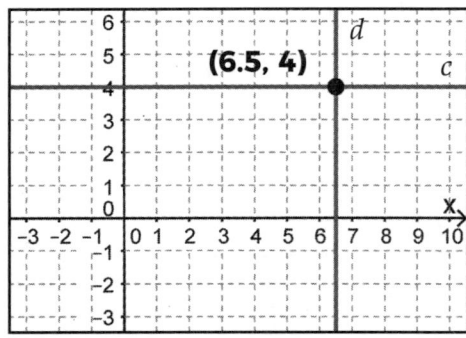

STEP 2 For line c, all of the y-values are 4, so write the equation of line c.

$y = 4$

STEP 3 For line d all of the x-values are 6.5, so write the equation of line d.

$x = 6.5$

STEP 4 Slope is the ratio of the vertical change to the horizontal change, $\frac{\Delta y}{\Delta x}$. Horizontal lines have a slope of 0 since Δy, the change in the vertical direction, is 0. Vertical lines have an undefined slope since Δx, the change in the horizontal direction, is 0 and the denominator of any fraction can never equal 0.

The slope of line c is 0 and the slope of line d is undefined.

PRACTICE

Write the equations of each of the following lines. Identify the slope, if it exists.

1. Line k is parallel to the y-axis, perpendicular to the x-axis, and passes through the point (−3, 6).

2. Line r is parallel to the y-axis and contains the point (4, −2).

3. Line t is perpendicular to the y-axis and passes through point (−3, 1.5).

Use the graph of line l to answer questions 4-8.   *Use the graph of line b to answer questions 9-11.*

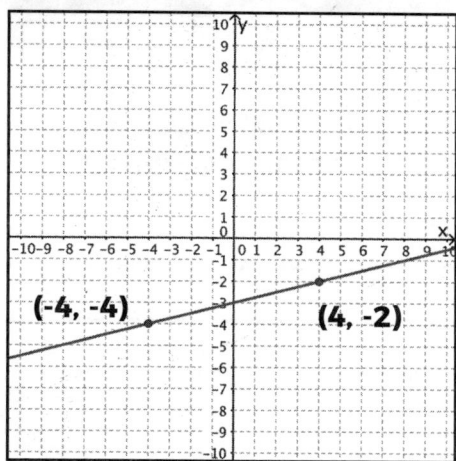

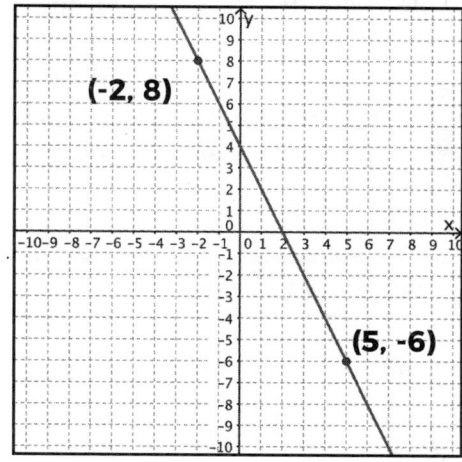

4. Write the equation, in slope-intercept form, of line *m* that is parallel to line *l* and has a *y*-intercept of (0, 6).

9. Which equation describes a line perpendicular to line *b* that has a *y*-intercept of (0, −2)?

A $y = 2x - 2$

B $y = \frac{1}{2}x - 2$

C $y = -2x - 2$

D $y = -\frac{1}{2}x - 2$

5. Write the equation, in standard form, of line *n* that is perpendicular to line *l* and passes through the point (3, −6).

10. Which equation describes a line parallel to line *b* that passes through the point (4, 3)?

F $y = -2x + 11$

G $y = 2x - 5$

H $y = \frac{1}{2}x + 1$

J $y = -\frac{1}{2}x + 5$

6. Write the equation of line *p* that is parallel to the *x*-axis and passes through line *l* at point (−8, −5).

7. Write the equation of line *q* that is perpendicular to the *x*-axis and passes through line *l* at point (4, −2).

11. What is the equation of line *j* that is perpendicular to the *y*-axis and passes through the *y*-intercept of line *b*?

A $y = 0$

B $x = 4$

C $y = 4$

D $x = 0$

8. Write the equation of the line that passes through (4, −2) and has a slope of 0.

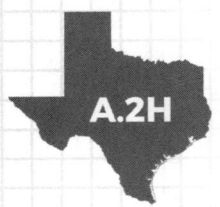

WRITING TWO-VARIABLE LINEAR INEQUALITIES

A.2H The student is expected to write linear inequalities in two variables given a table of values, a graph, and a verbal description.

ℹ️ TELL ME MORE...

An **inequality** shows a relationship between two quantities that are not equivalent or equal. In this case, one quantity is less than the other. You can use the symbols for **less than**, <, or **greater than**, >, to show these relationships.

The graph shows the inequality $y < 2x + 3$. The **solution set** of this inequality is the set of all ordered pairs (x, y) that when substituted into the inequality make a true statement.

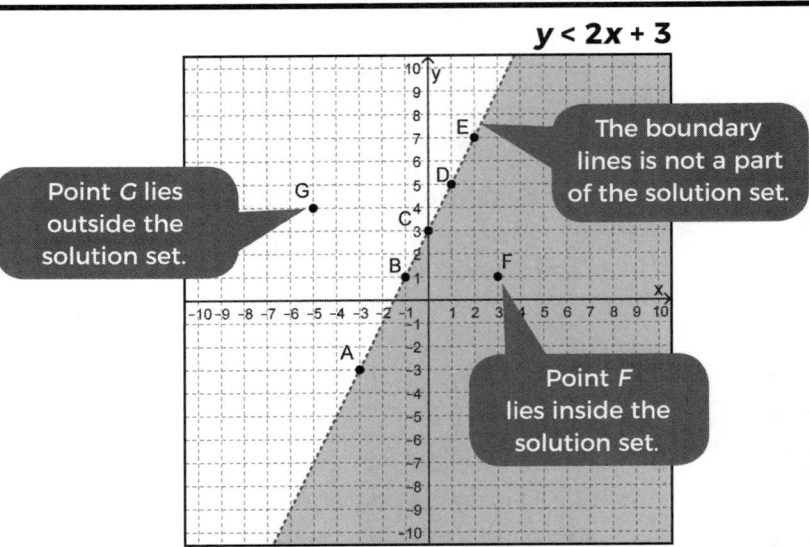

$y < 2x + 3$

Point G lies outside the solution set.

The boundary lines is not a part of the solution set.

Point F lies inside the solution set.

Points A, B, C, D, and E all lie on the **boundary line** of the inequality. Since the inequality reads "less than" and does not include "or equal to," the boundary line, including these five points, is not a part of the solution set.

Point A: $(-3, -3)$
$$-3 < 2(-3) + 3$$
$$-3 < -3$$
This is not a true statement, since $-3 = -3$.

Point F, $(3, 1)$ along with all other points in the shaded region, is a part of the solution set.
$$3 < 2(1) + 3$$
$3 < 5$, which is a true statement.

Point G, $(-5, 4)$, along with all other points in the *unshaded region*, is *not* a part of the solution set.
$$4 < 2(-5) + 3$$
$$4 < -7$$
This not is a true statement, since $4 > -7$.

In a graph, a solid boundary line indicates that the points on the line are included in the solution set. Since this is the graph of a line, then the inequality will include "or equal to:" ≤ or ≥. If the boundary line is dashed, then those points are not included in the solution set and the inequality will be < or >.

✓ EXAMPLES

EXAMPLE 1: Lone Star Community Theater sells tickets to one show as a fundraiser. Adult tickets cost $15 and student/senior tickets receive a 40% discount. Write an inequality that represents all possible combinations of a, the number of adult tickets, and s, the number of student/senior tickets that must be sold in order to meet or exceed the fundraising goal of $1000.

STEP 1 Determine the cost of one student/senior ticket. A 40% discount means that the sale price is 60% of the cost of one adult ticket.
$$0.6(\$15) = \$9$$
One student/senior ticket costs $9.

STEP 2 Use the price per ticket as a rate (dollars per ticket) to determine the money raised from selling a adult tickets and s student/senior tickets.

$$\text{Number of Tickets} \times \frac{\text{Cost (dollars)}}{\text{ticket}} \qquad \text{Number of Tickets} \times \frac{\text{Cost (dollars)}}{\text{ticket}}$$

$$a \times \$15 \qquad\qquad\qquad s \times \$9$$

$$15a \qquad\qquad\qquad 9s$$

Adult tickets: 15a dollars; Student/senior tickets: 9s dollars

STEP 3 Identify the inequality from the context of the problem.

Write an inequality that represents all possible combinations of a, the number of adult tickets, and s, the number of student/senior tickets that must be sold in order to <u>meet or exceed</u> the fundraising goal of $1000.

The word "meet" means equal. The word "exceed" means greater than.

The inequality will be greater than or equal, ≥.

STEP 4 Write the inequality by combining the money from each ticket type, the inequality symbol, and the fundraising goal.

$$\text{Money from Adult Tickets} + \text{Money from Student/Senior Tickets} \; \mathbf{?} \; \text{Fundraising Goal}$$
$$15a \qquad\qquad + \qquad\qquad 9s \qquad\qquad\qquad \geq \qquad\qquad 1000$$

15a + 9s ≥ 1000

EXAMPLE 2: The table shows points that lie on the boundary line that is included in the solution set of an inequality. The point (1, 3) is in the solution set of the inequality. Write the inequality relating x and y in slope-intercept form.

x	y
-2	-7.5
0	-5
2	-2.5
3	-1.25
4	0

STEP 1 Determine the slope of the boundary line from the table.

x	y
-2	-7.5
0	-5
2	-2.5
3	-1.25
4	0

$0 - (-2) = 2$ $\qquad$ $-5 - (-7.5) = 2.5$

$$m = \frac{\Delta y}{\Delta x} = \frac{2.5}{2} = \frac{5}{4}$$

$$m = \frac{5}{4}$$

STEP 2 Identify the y-coordinate of the y-intercept, or the function value when $x = 0$. This value will be the value of b.

$$b = -5$$

x	y
-2	-7.5
0	-5
2	-2.5
3	-1.25
4	0

STEP 3 Use the values of m and b to write the equation of the boundary line in slope-intercept form, $y = mx + b$.

$$m = \frac{5}{4} \text{ and } b = -5, \text{ so } y = \frac{5}{4}x - 5$$

STEP 4 Determine the inequality symbol that should be used. Use the point from the solution set, (1, 3) to determine if the inequality is greater than or less than. The boundary line is included in the solution set, so the inequality will include "or equal to."

$$y \geq \frac{5}{4}x - 5$$

$$y \; ? \; \frac{5}{4}x - 5$$

$$3 \; ? \; \frac{5}{4}(1) - 5$$

$$3 \; ? \; \frac{5}{4} - 5$$

$$3 > -\frac{15}{4}$$

EXAMPLE 3 Write the inequality with the solution set shown in the graph.

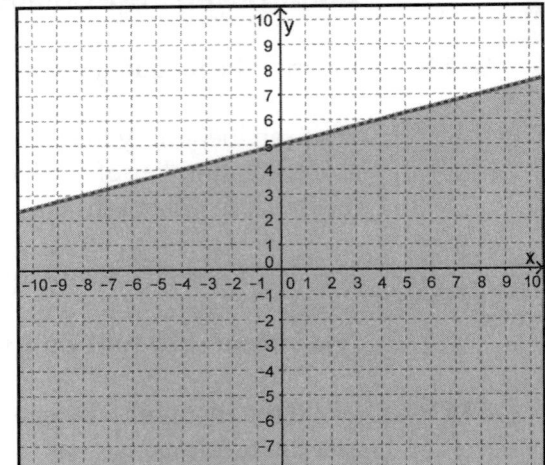

STEP 1 Determine the slope, m, and y-coordinate of the y-intercept, b, of the boundary line from the graph. Use this information to write the equation of the boundary line.

$$m = \frac{\Delta y}{\Delta x} = \frac{1}{4}$$

$$m = \frac{1}{4}, \; b = 5, \text{ so } y = \frac{1}{4}x + 5.$$

STEP 2 Determine the inequality symbol using any point from the solution set (shaded region), such as (0, 0).

The inequality is less than, <.

$$y = \frac{1}{4}x + 5$$

$$0 \; ? \; \frac{1}{4}(0) + 5$$

$$0 < 5$$

STEP 3 Write the inequality. The boundary line is dashed so it is not included in the solution set and "or equal to" will not be a part of the inequality.

$$y = \frac{1}{4}x + 5$$

PRACTICE

1. Clark has an orchard and grows peaches. There are 11 rows of peach trees in the orchard and in the past, the orchard has yielded no less than f tons of fruit. Each worker picks peaches from one row of trees. What inequality can be used to determine t, the approximate weight in tons of the peaches each worker will pick?

2. A taxi ride with Company A costs $2.25 plus $2.00 per mile. The taxi company also charges a fee of $0.40 for each minute spent waiting in traffic. If Markus needs his taxi ride to be less than $40, what inequality can be used to determine the number of miles for the taxi trip, x, and the number of minutes waiting in traffic, y, for Markus' taxi ride?

3. Thai needs to rent a truck to move some furniture. The table shows the maximum cost of the rental truck, y, based on the number of miles driven, x. What inequality describes the relationship shown?

x	y
0	30
16	34
20	35
44	41
50	42.5

A $y \leq 0.25x + 30$

B $y < 0.25x + 30$

C $y > 0.25x + 30$

D $y \geq 0.25x + 30$

5. The weight limit on the elevator in Joe's building is marked as 1,500 pounds. The table below shows combinations of males, m, and females, f, that can share the elevator at the weight limit based on their average weights.

m	f
1	11
3	8
5	5
7	2

The elevator could also hold 4 males and 5 females. What inequality written in standard form describes how many males and females can share the elevator within its weight limit in Joe's building

4. Rick's Rentals uses the graph to show the possible numbers of video games, x, and movies, y, that can be rented for less than $40 per month.

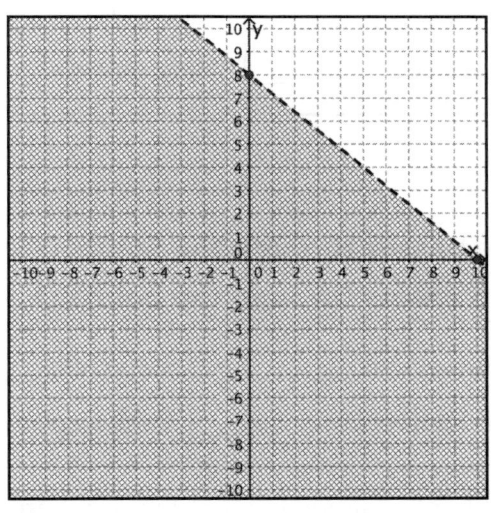

Write an inequality in standard form that represents this relationship.

6. The graph below shows an inequality. Which inequality statement best matches the graph?

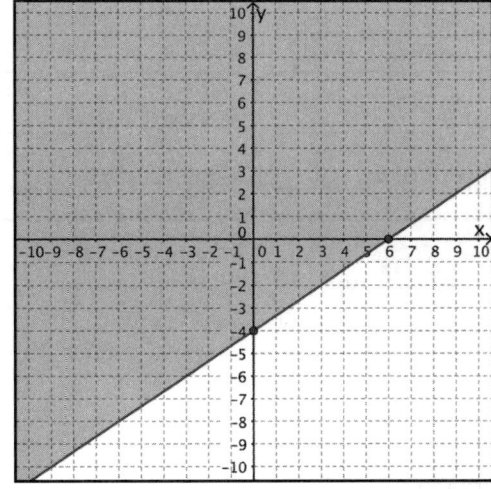

A $2x - 3y \geq 12$

B $2x + 3y \leq 12$

C $2x - 3y \leq 12$

D $2x + 3y \geq -1$

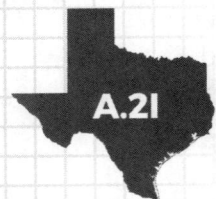

A.2I

The student is expected to write systems of two linear equations given a table of values, a graph, and a verbal description.

(i) TELL ME MORE...

Systems of equations are two or more equations that describe a situation. For two variables, you can generally write a system of two linear equations that describe the constraints on the variables for a given problem. Once you have a system of two linear equations, then you can solve the system and determine a solution. This lesson focuses on writing the system of two linear equations. Other lessons will show you how to solve the system.

To write a system of linear equations, you write one linear equation at a time. Use a verbal description, table of values, or graph to determine the slope or the coordinates of a point. The graph, table, and sentence below all describe the same line. Use the information presented to write the slope and y-intercept in the space provided.

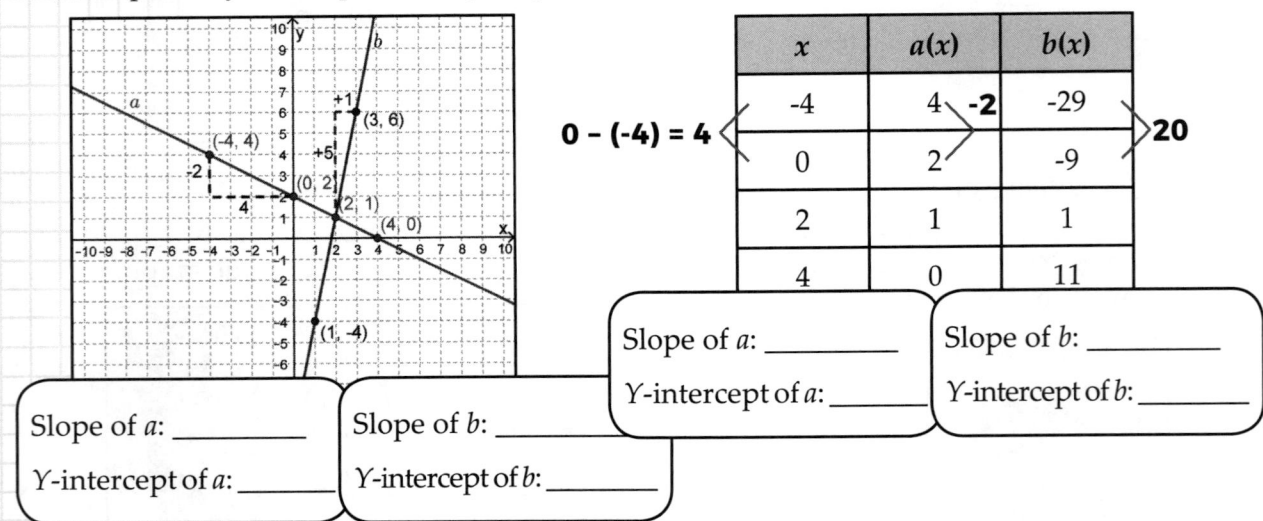

Slope of *a*: _____

Y-intercept of *a*: _____

Slope of *b*: _____

Y-intercept of *b*: _____

Slope of *a*: _____

Y-intercept of *a*: _____

Slope of *b*: _____

Y-intercept of *b*: _____

Line *a* passes through the points (–4, 4) and (0, 2).

Line *b* passes through the points (1, –4) and (2, 1).

Slope of *a*: _____

Y-intercept of *a*: _____

Slope of *b*: _____

Y-intercept of *b*: _____

Knowing the slope and one point on a line is enough information to write the equation of the line. For a system of linear equations, you need information for both lines. In a real-world context, you are given this information in a problem situation and will need to translate the words to algebra.

EXAMPLES

EXAMPLE 1: The graph of lines z_1 and z_2 are shown. Write a system of equations that represents the lines shown on this graph.

STEP 1 Identify two points on the graph of each line.

- z_1 appears to pass through (0, 3) and (2, –3).

- z_2 appears to pass through (0, –7) and (2, –3).

z_1: (0, 3) and (2, –3)

z_2: (0, –7) and (2, –3)

STEP 2 Count the slope in the graph of each line using Δx, the change in x between the two points, and Δy, the change in y between the two points.

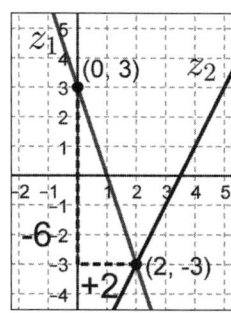
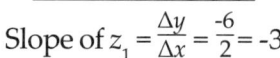
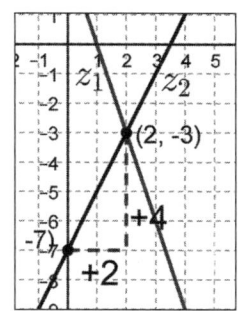

$$\text{Slope of } z_1 = \frac{\Delta y}{\Delta x} = \frac{-6}{2} = -3 \qquad \text{Slope of } z_2 = \frac{\Delta y}{\Delta x} = \frac{4}{2} = 2$$

z_1: m = –3

z_2: m = 2

STEP 3 Substitute the slope for m and the y-coordinate of the y-intercept for b into slope-intercept form.

$$y = mx + b \qquad y = mx + b$$
$$y = -3x + 3 \qquad y = 2x - 7$$

z_1: y = –3x + 3

z_2: y = 2x – 7

EXAMPLE 2: The freshmen class at Hurricane High School plans to sell flowers as a fundraiser to help victims of a recent storm. The table shows the cost to order x flowers at two local floral supply stores. Write a system of equations for $p(x)$ and $f(x)$.

STEP 1 A table contains sets of points for each function. Select 2 points (pairs of x- and y-values) for both $p(x)$ and $f(x)$.

$p(x)$: (10, 80) and (20, 110)

$f(x)$: (10, 50) and (20, 100)

Number of Flowers, x	Petal Pushers, $p(x)$	Flower Fans, $f(x)$
10	80	50
20	110	100
30	140	150
40	170	200
50	200	250

STEP 2 Calculate the slope of the line for each function using the slope formula.

$$p(x) : m = \frac{y_2 - y_1}{x_2 - x_1} = \frac{110 - 80}{20 - 10} = \frac{30}{10} = 3$$

$$f(x) : m = \frac{y_2 - y_1}{x_2 - x_1} = \frac{100 - 50}{20 - 10} = \frac{50}{10} = 5$$

- For $p(x)$, let $(10, 80) = (x_1, y_1)$ and $(20, 110) = (x_2, y_2)$.
- For $f(x)$, let $(10, 50) = (x_1, y_1)$ and $(20, 100) = (x_2, y_2)$.

$p(x)$: $m = 3$ and $f(x)$: $m = 5$

STEP 3 None of the points in the table represent an intercept, so use the point-slope formula instead of slope-intercept form.

$$y - y_1 = m(x - x_1) \qquad y - y_1 = m(x - x_1)$$
$$y - 80 = 3(x - 10) \qquad y - 50 = 5(x - 10)$$

- For $p(x)$, substitute $m = 3$, $x_1 = 10$, and $y_1 = 80$ into the general point-slope form.
- For $f(x)$, substitute $m = 5$, $x_1 = 10$, and $y_1 = 50$ into the general point-slope form.

$p(x)$: $y - 80 = 3(x - 10)$
$f(x)$: $y - 50 = 5(x - 10)$

STEP 4 Simplify each expression.

$p(x) = 3x + 50$
$f(x) = 5x$

$$y - 80 = 3(x - 10) \qquad y - 50 = 5(x - 10)$$
$$y = 3(x - 10) + 80 \qquad y = 5(x - 10) + 50$$
$$y = 3x - 30 + 80 \qquad y = 5x - 50 + 50$$
$$y = 3x + 50 \qquad\qquad y = 5x$$

EXAMPLE 3 A baseball team charges $25 for tickets for upper deck seats and $40 for tickets for lower deck seats. The stadium has 7500 seats and when the game is completely sold out (i.e., tickets for all seats have been sold), there is $255,000 in ticket sales. Write a system of equations that could be used to determine U, the number of upper deck seats and L, the number of lower deck seats that are in the stadium.

STEP 1 Determine the information that is given in the problem. Highlight the information if desired.

- Upper deck seat tickets cost $25 each and U represents the number of upper deck seats.
- Lower deck seat tickets costs $40 each and L represents the number of lower deck seats.

STEP 2 Write an expression describing the revenue generated by selling U upper deck seat tickets. Write an expression describing the revenue generated by selling L lower deck seat tickets.

Cost per Ticket × Number of Tickets Sold = Revenue

$25 × U = 25U$ for the revenue from upper deck seat tickets

$40 × L = 40L$ for the revenue from lower deck seat tickets

$25U$ and $40L$

STEP 3 Combine these costs to write a linear equation in standard form showing the amount of revenue generated.

Revenue from Upper Deck Tickets + Revenue from Lower Deck Tickets = Total Revenue

$$25U + 40L = 255{,}000$$

$25U + 40L = 255{,}000$

STEP 4 Write an equation showing how the number of upper deck seats and number of lower deck seats combine for the total seating in the stadium.

Number of Upper Deck Seats + Number of Lower Deck Seats = Total Seating

$$U + L = 7500$$

$U + L = 7500$
The system of equations is:
$25U + 40L = 255{,}000$
$U + L = 7500$

PRACTICE

1. The graph of lines p and q are shown.

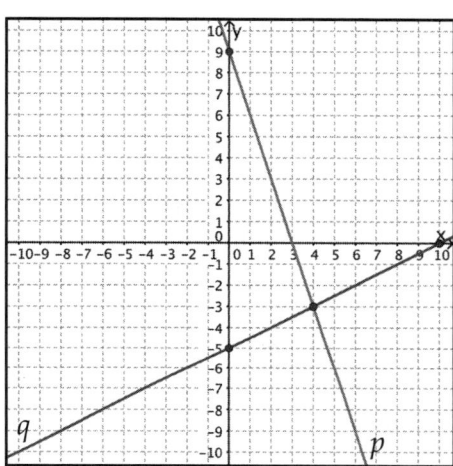

What system of linear equations, written in standard form, is best represented by the graph?

2. The table shows points that lie on $m(x)$ and $n(x)$. Both functions form a system of linear equations.

x	$m(x)$	$n(x)$
0	-5	40
1	-2.5	35
2	0	30
3	2.5	25
4	5	20

What system of linear equations do the points in the table model?

3. Damian has been saving coins he finds on the ground for several months. So far he has found 37 coins that are either pennies or nickels. The combined value of his found pennies and nickels currently is $1.09. What system of linear equations, written using standard form, represents the situation if Damian found p pennies and n nickels?

4. The sum of two numbers is 84. Twice the value of the first number less the second number is 21. What system of linear equations can be used to determine the two numbers, a and b?

5. The perimeter of a rectangular dog pen is 126 feet. The pen's length, l, is 9 feet less than 3 times its width, w. What system of linear equations can be used to determine the dimensions, in feet, of the rectangular dog pen?

6. Mary's Yoga Barn charges $30 for an annual membership plus $2.50 per class taken. Yoga Retreat charges customers $5.00 per class, but has no membership fee. The table shows some costs for both yoga studios based on the number of classes taken.

Number of Classes, x	Yoga Barn Cost, $f(x)$	Yogo Retreat Cost, $g(x)$
0	$30	$0
2	$35	$10
4	$40	$20
6	$45	$30
8	$50	$40

What system of linear equations can be used to find the point where the cost for the same number of classes at both yoga studios is equal?

7. Road Builders, Inc. hired 15 workers for a road construction project. Based on their skills, some of the workers, x, earned $165 per day and the rest of the workers, y, earned $190 per day. The daily payroll for the project is $2,600. Which system of linear equations should the company financial officer use to represent this situation?

A $x + y = 2,600$
$165x + 190y = 15$

B $x + y = 15$
$165x + 190y = 2,600$

C $x = y + 15$
$190x + 165y = 2,600$

D $x + y = 2,600$
$165x + 190y = 2,600$

8. Mrs. Grimm's science class recently took a field trip to the Exploratorium. Tickets to the Exploratorium cost $8.50 for students, s, and $14.25 for adults, a. When Mrs. Grimm turned in her paperwork for the school she reported buying a total of 27 tickets for a cost of $252.50. What system of linear equations can be used to determine the number of adults, a, and the number of students, s, that attended the field trip?

9. The graph shows the height of two hot air balloons over time. Based on the graph, which system of linear equations can be used to find the point at which both balloons are at the same height in the air at the same time?

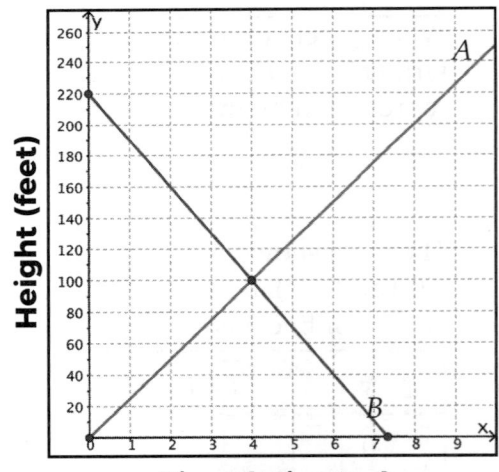

F $30x - y = 220$
$25x - y = 0$

G $30x + y = 220$
$25x - y = 0$

H $30x + y = -220$
$25x + y = 0$

J $30x - y = 220$
$25x + y = 0$

SOLVING LINEAR EQUATIONS

A.5A The student is expected to solve linear equations in one variable, including those for which the application of the distributive property is necessary and for which variables are included on both sides.

ℹ TELL ME MORE...

A **function** is a relationship between two variables: an **independent variable**, also called an input value, and a **dependent variable**, also called an output value.

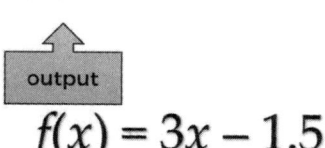

$$f(x) = 3x - 1.5$$

$$f(x) = 3x - 1.5$$

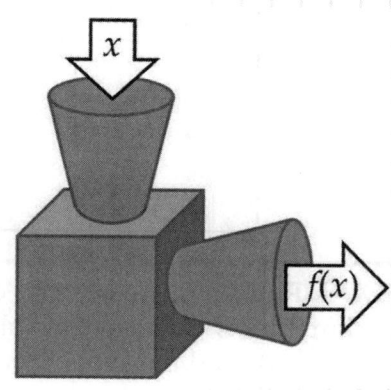

If you know the value of the independent variable, you can use that to evaluate the function. For example, if $f(x) = 3x - 1.5$, and you know the input value $x = 10$, then you can evaluate $f(10) = 3(10) - 1.5 = 28.5$.

Sometimes you know the output value and want to determine the input value that generates that output value. In that case, you use the output value to write an equation related to the function. Then, you use the properties of algebra to solve the equation. For example, if you know that $f(x) = 50$, you can write the equation $50 = 3x - 1.5$ and then solve for x using the properties of algebra.

✔ EXAMPLES

EXAMPLE 1: What is the solution to $-5x - 4(6.25x - 9) = 2(5x - 7)$?

STEP 1 To solve the equation for x, you need to have x by itself on one side of the equal sign and a number on the other. First, apply the distributive property to both sides of the equation.

–5x – 25x + 36 = 10x – 14

$$-5x - 4(6.25x - 9) = 2(5x - 7)$$
$$-5x - 4(6.25x) - 4(-9) = 2(5x) + 2(-7)$$
$$-5x - 25x + 36 = 10x - 14$$

STEP 2 Combine like terms.

–30x + 36 = 10x – 14

$$-5x - 25x + 36 = 10x - 14$$
$$-30x + 36 = 10x - 14$$

STEP 3 Apply the additive inverse to subtract $10x$ and 36 from both sides of the equation.

–40x + 36 = –14

$$-30x + 36 = 10x - 14$$
$$\underline{-10x -10x}$$
$$-40x + 36 = -14$$

STEP 4 Apply the additive inverse to subtract 36 from both sides of the equation.

–40x = –50

$$-40x + 36 = -14$$
$$\underline{-36 -36}$$
$$-40x = -50$$

STEP 5 Apply the multiplicative inverse to divide both sides of the equation by –40.

x = 1.25

$$-40x = -50$$
$$-40x \div -40 = -50 \div -40$$
$$x = 1.25$$

STEP 6 Use graphing technology to verify your answer. Treat each side (member) of the equation as one of a system of equations. Place the left member of the original equation into $Y1$ of the function editor and the right member into $Y2$ of the function editor. Look at the graph and calculate the intersection point. The x-coordinate of the intersection point is the solution to the original equation

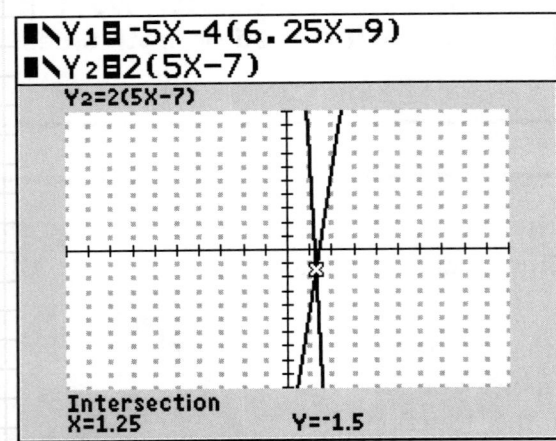

```
■\Y₁=-5X-4(6.25X-9)
■\Y₂=2(5X-7)
Y2=2(5X-7)



Intersection
X=1.25          Y=-1.5
```

EXAMPLE 2: A used car lot contains passenger cars and pickup trucks. The average value of a passenger car on the lot is $6,500 and the average value of a pickup truck on the lot is $8,000. There are 20 vehicles on the lot. The equation below can be used to determine t, the number of pickup trucks on the lot if there are c passenger cars.

$$t = \frac{142,000 - 6500c}{8000}$$

If there are 8 pickup trucks on the lot, how many passenger cars are there on the lot?

STEP 1 t represents the number of pickup trucks so $t = 8$. Substitute $t = 8$ into the equation.

$$8 = \frac{142,000 - 6500c}{8000}$$

STEP 2 Solve for c using the properties of algebra.

$$8 = \frac{142,000 - 6500c}{8000}$$

$$8(8000) = \left(\frac{142,000 - 6500c}{8000}\right)(8000)$$

$$64,000 = 142,000 - 6500c$$

$$\underline{-142,000 \quad -142,000}$$

$$-78,000 = -6500c$$

$$\frac{-78,000}{-6500} = \frac{-6500c}{-6500}$$

$$12 = c$$

$c = 12$, so there are 12 passenger cars on the lot.

EXAMPLE 3: Luke worked 12 hours and earned $16 per hour. Luke earned $8.25 more than his brother, Cooper, who worked 15 hours. What is Cooper's hourly pay in dollars and cents? Record your answer and fill in the bubbles on your answer document.

STEP 1 Write an equation to determine c, Cooper's hourly pay in dollars and cents.

- Luke worked 12 hours at $16 per hour. $12(16) = \$192$.

- Cooper worked 15 hours at c dollars per hour. $15c$

- Luke earned $8.25 more than Cooper, so Luke = Cooper + $8.25

192 = 15c + 8.25

STEP 2 Use the properties of algebra to solve for c.

c = 12.25, so Cooper earned $12.25 per hour.

$$192 = 15c + 8.25$$
$$192 - 8.25 = 15c + 8.25 - 8.25$$
$$183.75 = 15c$$
$$\frac{183.75}{15} = \frac{15c}{15}$$
$$12.25 = c$$

STEP 3 Since the question is a gridded response question, enter your response on the grid provided. Practice using the grid with the instructions.

1. Record a 1 in the first column containing numbers. Record a 2 in the next column. Record a decimal in the next column. Record a 2 in the next column. Record a 5 in the next column.

2. Bubble the **1** beneath the numeral 1. Bubble **2** beneath the numeral 2. Bubble **.** beneath the decimal. Bubble **2** beneath the numeral 2. Bubble **5** beneath the numeral 5.

PRACTICE

Answer the questions that follow.

1. What is the solution to the following equation?

$$\tfrac{1}{3}(15x - 21) + 6 = \tfrac{3}{4}(20 + 32x) - 9x$$

2. What value of x makes the equation $3x - (7 - x) = \tfrac{1}{2}(5x + 4)$ true?

3. If $y = 4(2x - 3) + 7 - (3x + 7.5)$ what is the value of x when $y = 17.5$?

4. The formula below can be used to convert temperature in degrees Fahrenheit, F, to temperatures in degrees Celsius, C.

$$C = \frac{5}{9}(F - 32)$$

On a certain day at the North Pole the thermometer reads $-25°C$. What is the temperature in degrees Fahrenheit for that day at the North Pole? Record your answer and fill in the bubbles on your answer document.

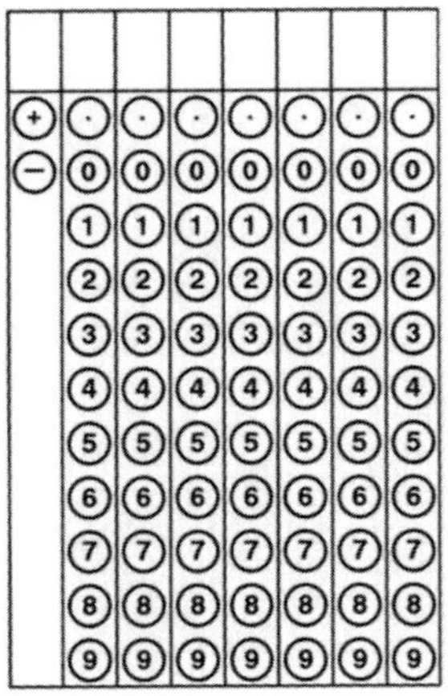

5. In a certain isosceles triangle, the measures of each of the two congruent angles is 15° less than one-half the measure of the third angle. What are the measures of the three angles?

6. The perimeter of a rectangular deck is 78 feet. The length of the deck can be represented as $(2x + 3)$, and its width can be represented as $(3x - 4)$. What are the dimensions of the rectangular deck in feet?

7. Beto wanted to find the value of three consecutive odd numbers with a sum of 87. He used the equation shown below to solve for the numbers.

$$n + (n + 2) + (n + 4) = 87$$

What are the three numbers?

8. What value of x makes the equation $2 - 3(4x + 1) - x = 6 - (3x + 2)$ true?

A $x = -\frac{1}{2}$

B $x = -\frac{5}{2}$

C $x = -\frac{3}{10}$

D $x = -\frac{7}{2}$

9. Jason solved the equation below for the value of h.

Step 1: $4 - 6(3h + 2) = 1 - (5 - 7h)$

Step 2: $-2(3h + 2) = 1 - 5 - 7h$

Step 3: $-6h - 4 = -4 - 7h$

Step 4: $h = 0$

Which statement about Jason's solution is true?

F Jason's solution is correct.

G Jason's solution is incorrect because he made an error in step 2.

H Jason's solution is incorrect because he made 2 errors in step 2.

J Jason's solution is incorrect because he made an error in step 4.

SOLVING LINEAR INEQUALITIES

A.5B

The student is expected to solve linear inequalities in one variable, including those for which the application of the distributive property is necessary and for which variables are included on both sides.

ⓘ TELL ME MORE...

Linear inequalities are related to linear equations. In an equation, both sides of the equation are equivalent. So, in a linear equation with one variable, there is only one possible value for x. On the other hand, both sides of a linear inequality may not be equivalent. One side of the inequality is "less than" or "greater than" the other. Hence, instead of using an equal sign, we use inequality symbols < and > to show the relationship between the two quantities.

$3x - 5 = 10$	$3x - 5 < 10$	$3x - 5 > 10$
$x = 5$	$x < 5$	$x > 5$

Inequalities involve a range of possible solutions with defined endpoints.

- If that endpoint is not included in the solution set, then the inequality symbol will not include "or equal to:" > or <. Graphically, it is represented with an open circle.

- If that endpoint is included in the solution set, then the inequality symbol will include "or equal to:" ≥ or ≤. Graphically, it is represented with a closed circle.

When solving inequalities symbolically, remember that a negative coefficient means "the opposite of" the number. So if you multiply or divide by that negative coefficient, then you must also take "the opposite of" the inequality symbol and reverse it, as in Example 2.

✓ EXAMPLES

EXAMPLE 1: Write an inequality that describes all the solutions to $2.5(3x - 11) < -2.5x + 6$. Represent the solution on a number line.

STEP 1 Use the properties of algebra to obtain an inequality with x alone on one side of the inequality symbol and a number on the other. Begin by distributing 2.5 on the left side of the inequality.

$$2.5(3x - 11) < -2.5x + 6$$
$$2.5(3x) + 2.5(-11) < -2.5x + 6$$
$$7.5x - 27.5 < -2.5x + 6$$

7.5x – 27.5 < –2.5x + 6

STEP 2 Add 2.5x to both sides of the inequality so that the x term is on the left.

$$7.5x - 27.5 < -2.5x + 6$$
$$\underline{+ 2.5x \qquad\qquad + 2.5x}$$
$$10x - 27.5 < 6$$

10x – 27.5 < 6

STEP 3 Add 27.5 to both sides of the inequality so that the x term has no addend.

$$10x - 27.5 < 6$$
$$+ 27.5 + 27.5$$
$$\overline{10x < 33.5}$$

10x < 33.5

STEP 4 Divide both sides by 10.

$$\frac{10x}{10} < \frac{33.5}{10}$$
$$x < 3.35$$

x < 3.35

STEP 5 Represent the solution, $x > 3.35$, on a number line. Plot 3.35 and then draw an arrow toward all values greater than 3.35.

EXAMPLE 2: Write an inequality that describes all the solutions to $5(x - 7.5) \geq 2(5x + 9)$.

STEP 1 Use the properties of algebra to obtain an inequality with x alone on one side of the inequality symbol and a number on the other. Begin by distributing 5 on the left side of the inequality and 2 on the right.

5x – 37.5 ≥ 10x + 18

STEP 2 Subtract $10x$ from both sides of the inequality so that the x term is on the left.

$$5x - 37.5 \geq 10x + 18$$
$$-10x -10x$$
$$\overline{-5x - 37.5 \geq 18}$$

–5x – 37.5 ≥ 18

STEP 3 Add 37.5 to both sides of the inequality so that the x term has no addend.

$$-5x - 37.5 \geq 18$$
$$+37.5 + 37.5$$
$$\overline{-5x \geq 55.5}$$

–5x ≥ 55.5

STEP 4 Divide both sides by –5 and reverse the inequality symbol.

$$\frac{-5x}{-5} \geq \frac{55.5}{-5}$$
$$x \leq 11.1$$

x ≤ 11.1

> ## YOU TRY IT!
>
> Solve $1.8(2x - 10) > -6(1.5x - 9)$
>
> Distribute 1.8 on the left and –6 on the right.
>
> Move one x term to the other side (add its opposite).
>
> Move the constant term that is an addend to the only x term (add its opposite).
>
> Divide by the coefficient of x (do you need to reverse the symbol?)
>
> Inequality: _____

EXAMPLE 3: Betsy's Bayside Barbeque is having a fundraiser for the local animal shelter. For an all-you-can-eat barbeque, they charge $15 for a child's ticket and $35 for an adult's ticket. Betsy has a goal of raising $1200. If 20 child's tickets are sold, what is the minimum amount of adult's tickets that must be sold in order to reach the fundraising goal?

STEP 1 Write a linear inequality in one variable for this situation. Let a represent the number of adult's tickets and c represent the number of child's tickets.

- Each child's ticket costs $15, so if c child's tickets are sold, then that raises $15c$.

- We know that 20 child's tickets were sold, so $15(20) = 300.

- Each adult's ticket costs $35, so if a adult's tickets are sold, then that raises $35a$.

- The combination of child's tickets and adult's tickets must be greater than or equal to $1200 in order to meet the fundraising goal.

$$\text{Funds from Child's Tickets} + \text{Funds from Adult's Tickets} \geq \text{Goal}$$
$$\$300 + 35a \geq \$1200$$

300 + 35a ≥ 1200

STEP 2 Solve the inequality. Begin by adding −300 (subtracting 300) to both sides of the inequality.

35a ≥ 900

$$\begin{aligned} 300 + 35a &\geq 1200 \\ -300 \qquad &\quad -300 \\ 35a &\geq 900 \end{aligned}$$

STEP 3 Divide both sides by 35.

- You can only sell a whole number of tickets, so round up to the next whole number.

$$\frac{35a}{35} \geq \frac{900}{35}$$
$$a \geq 25.71$$

a ≥ 25.71, so 26 adult tickets must be sold to meet the fundraising goal.

PRACTICE

1. What values of x make the following inequality true?

$$4(3x - 5) + 8.5 < 7x + 9$$

2. What inequality describes all the solutions to
$$-8x - 2(4 - x) \leq -5(2x - 4) + 12x?$$

3. Maria solved an inequality. Her work is shown below.

$$2x + 6 > 5x - 1$$
$$7x + 6 > -1$$
$$7x > -7$$
$$x > -1$$

What error did Maria make in solving the inequality?

4. The first two steps in the solution process of an inequality are solved below.

$$3x + 2 \geq -16$$
$$3x \geq -18$$

What is the next step in the solution process to solve the inequality?

5. The measure of an obtuse angle is given as $(4x - 36)°$. What inequality describes all the possible values of x for the angle?

6. The area of a triangular garden can be no more than 120 square feet. The base of the triangle is 16 feet in length. What inequality describes the possible heights, h, of the triangular garden?

7. Determine and graph the inequality that describes all of the solutions to $-7(x - 3) \geq 42$.

8. Marlon is renting a van to move. The rental company charges a daily rate of $25 for a van rental plus 15¢ per mile driven when rounded to the nearest mile. If Marlon budgeted $80 to spend on the van for the move, what is the maximum number of whole miles he can drive?

9. The solution set to an inequality is shown graphed on the number line below.

Which of the following inequalities is represented by the graphed solution?

A $2(6 - x) \geq -9x + x$

B $x - 16 > 8x - 2$

C $7x + 3x > -8 + 6x$

D $5x - 1 > 13 - 2x$

10. Carlos is choosing a new cell phone plan. He has been comparing two plans and both offer the same amount of included free minutes. Plan A costs $39.99 per month with additional minutes over the plan costing $0.45 each. Plan B costs $45.99 per month with additional minutes over the plan costing $0.40 each. What inequality represents the number of additional minutes, m, Carlos can talk and have Plan A stay less expensive than Plan B?

F $m < 1,719$

G $m < 120$

H $m < 101$

J $m < 6$

SOLVING SYSTEMS OF LINEAR EQUATIONS

A.5C The student is expected to solve systems of two linear equations with two variables for mathematical and real-world problems.

ⓘ TELL ME MORE...

A **system of equations** can be solved algebraically as well as graphically or using tables. There are many algebraic methods of solving a system of equations. For systems of linear equations, two common methods are substitution and elimination

SUBSTITUTION

Goal: Substitute one variable, in terms of the other, into one equation to solve for that variable. Then you can use the value of the known variable to solve for the other variable.

Best Uses: If you have at least one linear equation in slope-intercept form ($y = mx + b$) then you can substitute $mx + b$ for y in the second equation.

ELIMINATION

Goal: Add two equations together in such a way that you eliminate (add to 0) one variable. Then you can solve for the remaining variable and use that value to go back and solve for the variable you eliminated.

Best Uses: If you have two linear equations in standard form ($Ax + By = C$), you can often eliminate one variable.

✓ EXAMPLES

EXAMPLE 1: What is the solution to this system of equations?

$$5x - 8y = 20$$
$$y = -\frac{3}{4}x + 8.5$$

STEP 1 Determine a solution strategy: substitution or elimination.

- One equation is in standard form and one is in slope-intercept form.
- Since one equation is written in slope-intercept ($y = mx + b$) form, use substitution.

Use the substitution method.

STEP 2 Substitute the expression for y into the standard form equation, $5x - 8y = 20$.

$$5x - 8(-\frac{3}{4}x + 8.5") = 20$$

$$5x - 8(-\frac{3}{4}x + 8.5) = 20$$

STEP 3   Use the properties of algebra to solve for x.

$x = 8$

$$5x - 8(-\tfrac{3}{4}x + 8.5) = 20$$
$$5x - 8(-\tfrac{3}{4}x) - 8(8.5) = 20$$
$$5x + 6x - 68 = 20$$
$$11x = 88$$
$$x = 8$$

STEP 4   Determine the value of y by substituting $x = 8$ into either equation. Since one equation is in slope-intercept form, use that equation to simplify the algebra.

$y = 2.5$

$$y = -\tfrac{3}{4}x + 8.5$$
$$y = -\tfrac{3}{4}(8) + 8.5$$
$$y = -6 + 8.5$$
$$y = 2.5$$

STEP 5   Write the solution as an ordered pair, (x, y).

(8, 2.5)

EXAMPLE 2: A university runs parking shuttles from the student union to two different parking lots. Each route is a different length and makes a full loop back to the student union.

■ On Monday, a shuttle traveled the Red Line route 8 times and the Yellow Line route 5 times, traveling a total of 24.5 miles.

■ On Tuesday, a shuttle traveled the Red Line route 11 times and the Yellow Line route 7 times, traveling a total of 34 miles.

What is the length of the each route in miles?

STEP 1   Let r represent the length of the Red Line route and y represent the length of the Yellow Line route. Write a system of equations from the problem information.

■ Monday: $8r + 5y = 24.5$

■ Tuesday: $11r + 7y = 34$

$8r + 5y = 24.5$

$11r + 7y = 34$

STEP 2   Since both equations are in standard form, solve using elimination.

■ Multiply the first equation by 7 and the second equation by -5.

■ Add the resulting equations together to eliminate y.

$$8r + 5y = 24.5 \rightarrow 7(8r + 5y = 24.5) = 56r + 35y = 171.5$$
$$11r + 7y = 34 \rightarrow -5(11r + 7y = 34) = -55r - 35y = -170$$
$$1r + 0y = 1.5$$

$r = 1.5$

> **YOU TRY IT!**
>
> Solve the system.
>
> $$y = 1.5x - 9$$
> $$2x + 8y = 12$$
>
> Solution strategy: _____
>
> Solution:
>
> $(x, y) = $ _____

STEP 3 Determine the value of y by substituting $r = 1.5$ into either of the two equations.

$y = 2.5$

$$11r + 7y = 34$$
$$11(1.5) + 7y = 34$$
$$16.5 + 7y = 34$$
$$7y = 17.5$$
$$y = 2.5$$

STEP 4 Interpret the values of r and y in the problem situation.

- r represents the length of the Red Line route, so if $r = 1.5$, then the Red Line route is 1.5 miles long.

- y represents the length of the Yellow Line route, so if $y = 2.5$, then the Yellow Line route is 2.5 miles long.

The Red Line route is 1.5 miles long and the Yellow Line route is 2.5 miles long.

EXAMPLE 3: An art gallery displays both color and black-and-white photographs. There are 85 photographs on display and the number of color photographs is 11 fewer than twice the number of black-and-white photographs. How many black-and-white photographs are there? Record your answer and fill in the bubbles on your answer document.

STEP 1 Let b represent the number of black-and-white photographs and c represent the number of color photographs. Write a system of equations from the information in the problem.

- "There are 85 photographs on display" → color + black-and-white = 85, $c + b = 85$

- "the number of color photographs is 11 fewer than twice the number of black-and-white photographs" → color = 11 fewer than 2(black-and-white), $c = 2b - 11$

$c + b = 85$

$c = 2b - 11$

STEP 2 Determine a solution strategy. Since one equation is in $c =$ form, use substitution.

Substitution

STEP 3 Substitute the expression for c into the standard form equation.

$$c + b = 85$$
$$(2b - 11) + b = 85$$

$(2b - 11) + b = 85$

STEP 4 Use the properties of algebra to solve for b.

$$(2b - 11) + b = 85$$
$$3b - 11 = 85$$
$$3b = 96$$
$$b = 32$$

$b = 32$

STEP 5 Since the question is a gridded response question, enter your response on the grid provided. Practice using the grid with the instructions.

1. Record a 3 in the first column containing numbers. Record a 2 in the next column.

2. Bubble the **3** beneath the numeral 3. Bubble **2** beneath the numeral 2.

PRACTICE

1. A system of linear equations is graphed on the coordinate plane below.

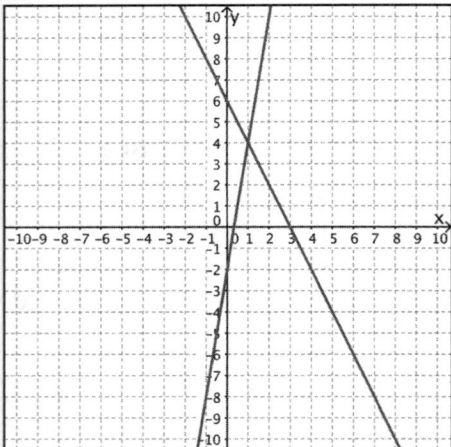

What is the solution to the system of equations?

2. Bill and Sam went to watch a basketball game together. While at the event, Bill bought 2 burger baskets and 3 sodas for $16.50. Sam bought 1 burger basket and 2 sodas for $9.50. What was the cost of a soda at the basketball game?

3. The tables below describe a system of linear equations.

x	y
-2	-7
-1	-6
0	-5
1	-4
2	-3

x	y
0	-13
1	-11
2	-9
3	-7
4	-5

What ordered pair represents the solution to the system?

4. The perimeter of a rectangular garden is 42 feet. The length, l, of the garden is 5 feet more than twice the width, w. What is the length of the garden?

5. Pema brought canned and packaged non-perishable food items for the food bank drive at school. The system of linear equations shown can be used to determine the number of cans and the number of packaged items Pema brought.

$$c + p = 30$$
$$c = 2p$$

How many of each type of food item did Pema donate for the food drive?

6. The equation $4x - 5y = 20$ is part of a system of linear equations. The second line in the system passes through the point (−5, 3). If the system formed by these two equations has no solution, what is the equation of the second line in the system?

7. What is the value of x in the solution to the system of linear equations below?

$$2x + 3y = 7$$
$$x + y = 3$$

Record your answer and fill in the bubbles on your answer document.

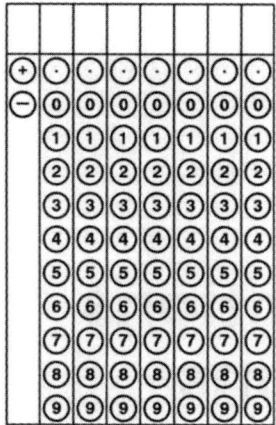

9. Xin is creating a snack mix using cheese crackers and pretzels she bought in bulk. The total weight of the snack items in the recipe is 40 ounces. The cheese crackers cost $0.25 per ounce and the pretzels cost $0.20 per ounce to purchase. Xin spent $8.70 on the snack items. How many ounces of cheese crackers are included in Xin's snack mix?

A 18 ounces

B 26 ounces

C 14 ounces

D 22 ounces

8. The drama class sold 924 tickets to the fall performance. They sold adult tickets for $8 each and student tickets for $5 each. They took in a total amount of $5,592 in ticket sales. The drama teacher wrote the following system of linear equations to represent the ticket sale information.

$$s + a = 924$$
$$5s + 8a = 5,592$$

How many of each type of ticket were sold for the fall performance?

10. What is the value of y in the solution this system of equations?

$$7x + 2y = 8$$
$$5x + 2y = 4$$

F 6

G 2

H −3

J −17

Unit 4
Quadratic Functions and Equations

DOMAIN AND RANGE OF QUADRATIC FUNCTIONS

A.6A The student is expected to determine the domain and range of quadratic functions and represent the domain and range using inequalities.

ⓘ TELL ME MORE...

The **domain** of a function is the set of input values that are used for the independent variable. The range of a function is the set of output values for the dependent variable. For any quadratic function, $f(x) = ax^2 + bx + c$, the domain is the set of all real numbers. The range, however, is bounded by the vertex of the graph of $f(x)$. Use the graph to identify the range of $f(x)$ and $g(x)$.

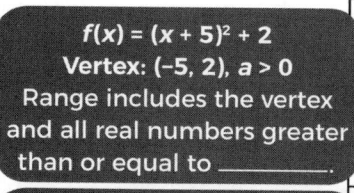

$f(x) = (x + 5)^2 + 2$
Vertex: (–5, 2), $a > 0$
Range includes the vertex and all real numbers greater than or equal to _____.

$g(x) = -(x - 4)^2 - 1$
Vertex: (4, –1), $a < 0$
Range includes the vertex and all real numbers less than or equal to _____.

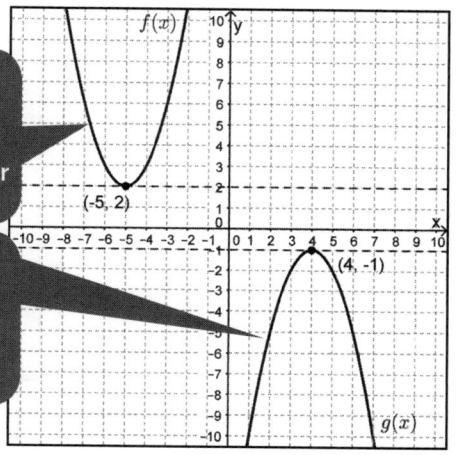

Domain

For any quadratic function, $f(x) = ax^2 + bx + c$, the domain is the set of all real numbers.

$$-\infty < x < \infty$$

Range

For any quadratic function, $f(x) = ax^2 + bx + c$, the range is the set of real numbers above or below the vertex, (h, k).

If $a > 0$: $f(x) \geq k$

If $a < 0$: $f(x) \leq k$

If the parabola opens upward, then the vertex is a minimum function value and the range is the set of real numbers greater than or equal to the y-coordinate of the vertex. If the parabola opens downward, then the vertex is a maximum function value and the range is the set of real numbers less than or equal to the y-coordinate of the vertex.

However, when a quadratic function is used to model real-world situations, then the domain and range can be restricted to numbers that make sense for the situation. In these cases, the domain or range can be written using inequalities for a **continuous** interval.

✓ EXAMPLES

EXAMPLE 1: What is the domain and range of $j(x) = 8 - (x - 3)^2$?

STEP 1 Graph the function to visually inspect the possible range.

- The parabola opens downward.

- The vertex appears to be (3, 8).

The possible range is $j(x) \leq 8$.

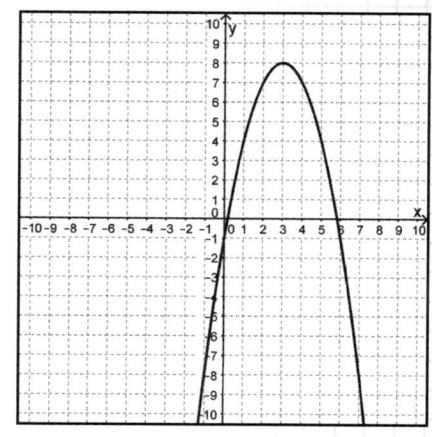

STEP 2 Use the commutative property of addition to rewrite $j(x)$ in vertex form, $\qquad j(x) = 8 - (x - 3)^2$
$j(x) = a(x - h)^2 + k$, in order to determine the coordinates of the vertex. The $\quad j(x) = -(x - 3)^2 + 8$
y-coordinate of the vertex will confirm the boundary of the range.

The vertex is (3, 8).

STEP 3 State the domain and range of the function.

$a < 0$, so the y-coordinate of the vertex, 8, is a maximum value.

The domain of $j(x)$ is all real numbers.
The range of $j(x)$ is all real numbers less than or equal to 8, or $j(x) \leq 8$.

EXAMPLE 2: While on vacation with his family, Carlos accidentally dropped a tennis ball from the balcony of their hotel room that is 144 feet above the ground. The function $h(t) = -16t^2 + 144$ describes the height, h, of the ball above the ground in feet t seconds after Carlos dropped it. In this situation, what is a reasonable domain for which the ball was in the air?

STEP 1 The domain of a quadratic function is all real numbers, but the situation may impose limits. Graph $h(t)$ to get a better idea of those limits.

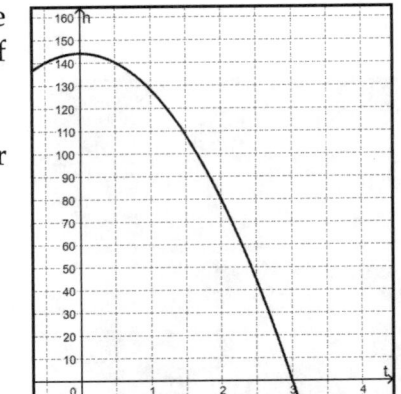

STEP 2 Determine any limits on the independent variable, time, or dependent variable, height above the ground.

- In this situation, it does not make sense for time, the independent variable, to be negative.

- In this situation, it does not make sense for the height above the ground, the dependent variable, to be negative.

Both variables are limited to positive numbers (Quadrant I) in this situation.

STEP 3 State the domain of the function for this situation, keeping in mind the constraints already observed.

- x can be any real number between 0 and 3, including both 0 and 3.

The domain of the function in this situation is $0 \leq t \leq 3$.

EXAMPLE 3 The graph of a part of quadratic function d is shown. Write the domain and range of the part shown using inequalities.

STEP 1 Determine the coordinates of each endpoint. There is a closed endpoint at $(-6, 0)$ and an open endpoint at $(1, -7)$.

(–6, 0) and (1, –7)

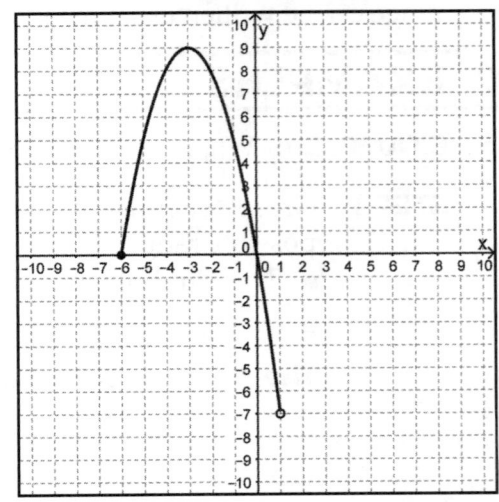

STEP 2 Identify the lower limit and upper limit for x using the x-coordinates of the endpoints. If the endpoint is an open endpoint (open circle), then that value is not included in the domain.

$$(-6, 0) \text{ and } (1, -7)$$

> –6 is less than 1, so it is the lower limit. The endpoint is a closed circle so it is included in the domain.

> 1 is greater than –6, so it is the upper limit. The endpoint is an open circle so it is not included in the domain.

–6 is the lower limit (included) and 1 is the upper limit (not included)

STEP 3 Write the domain using the lower limit, inequalities, and the upper limit. If the limit is included in the domain, be sure the inequality contains "or equal to."

$-6 \leq x < 1$

STEP 4 Locate the vertex on the graph and determine if it is a maximum or minimum value. Locate the endpoint that is farther from the vertex. The y-coordinates, or function values, of these points will determine the boundaries for the range of d.

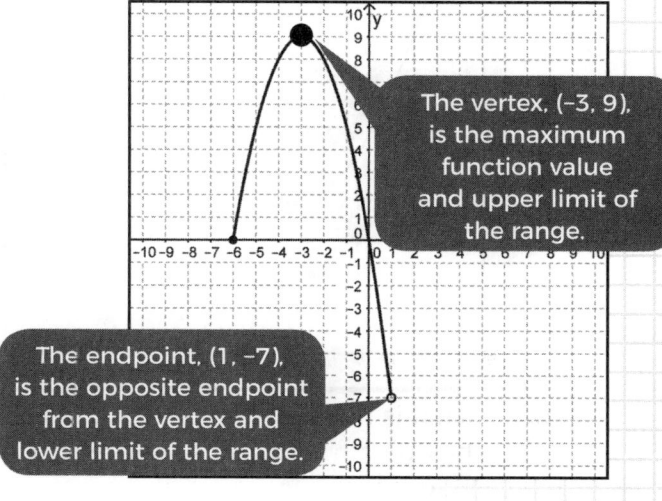

> The vertex, (–3, 9), is the maximum function value and upper limit of the range.

–7 is the lower limit (not included) and 9 is the upper limit (included)

> The endpoint, (1, –7), is the opposite endpoint from the vertex and lower limit of the range.

STEP 5 Write the range using the lower limit, inequalities, and the upper limit. If the limit is included in the range, be sure the inequality contains "or equal to."

$-7 < d(x) \leq 9$

EXAMPLE 4 The table shows some ordered pairs that belong to quadratic function q. What is the range of q?

x	$q(x)$
3	10
4	-2
5	-6
6	-2
7	10
8	30
9	58

STEP 1 Use the table to locate the vertex, which will reveal the limit of the range. The vertex is the point in the row in which the function values stop decreasing and begin increasing.

The vertex is (5, –6).

STEP 2 Determine if the vertex is a maximum or minimum function value.

As x increases from 3 to 5, $q(x)$ decreases. Then, as x increases from 5 to 9, $q(x)$ increases. So, at the vertex, (5, –6), the function values have decreased as low as they will go.

x	$q(x)$	
3	10	↓
4	-2	↓
5	-6	↓
6	-2	↑
7	10	↑
8	30	↑
9	58	↑

The vertex is a minimum function value.

STEP 3 Write the range of q using the vertex as the minimum function value.

$-6 \leq q(x)$

PRACTICE

Identify the domain and range for each function below using inequalities.

1. $f(x) = 2x^2 + 8x - 24$

3. $h(x) = -2.5(x + 1)^2$

2. $f(x)$, graphed below.

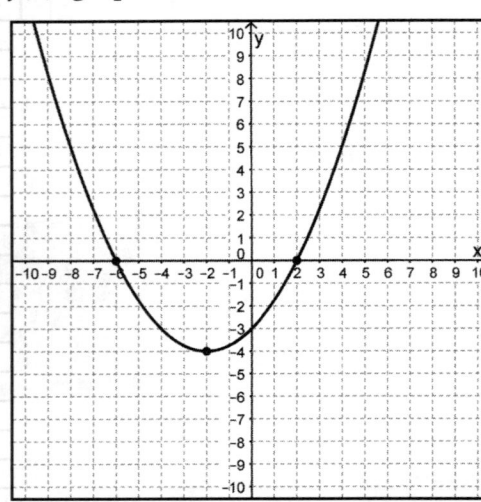

4. $k(x)$, graphed below.

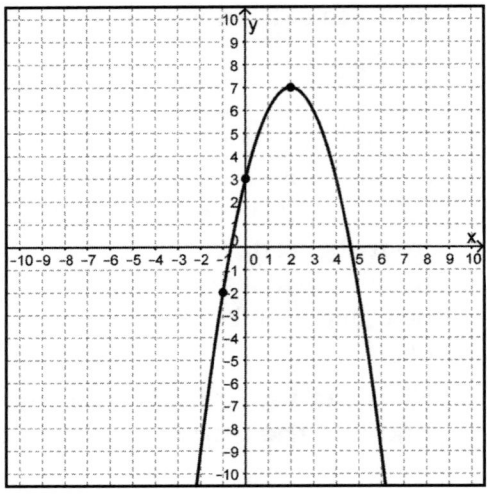

5. What is the range of $y = 12 - x^2$ when the domain is the set of integers from -3 to 2?

7. The path of a toy rocket is graphed below. The height of the rocket, $f(t)$, is a function of the number of seconds, x, the rocket is in the air after launch.

6. The table below shows points that belong to the function $g(x)$. What is the domain of $g(x)$? What is the range of $g(x)$ for the domain interval $-3 < x < 6$?

x	-5	0	3	7
$g(x)$	18	-12	-6	30

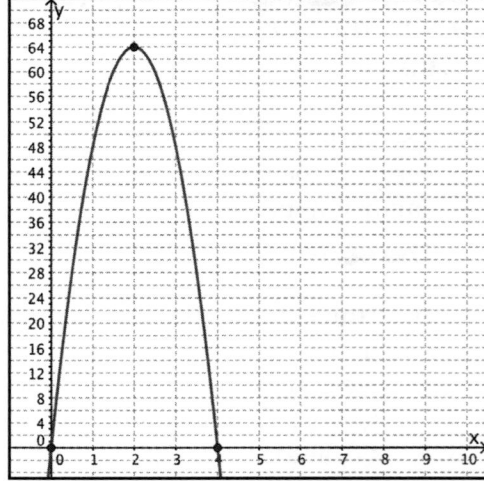

What is the domain of the graph of the rocket's path?

8. The graph of a function is shown below. Use an inequality to write the domain of the function.

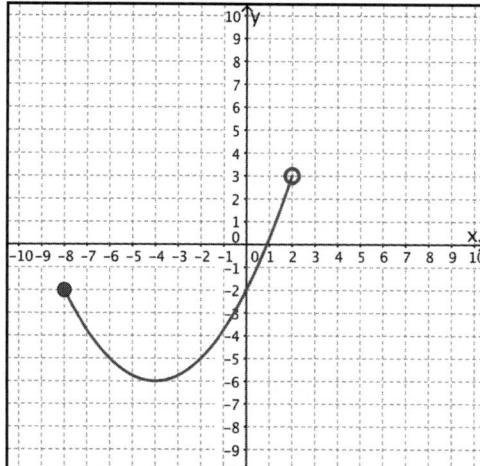

9. A rock is thrown into the air. The height of the rock in meters, $h(t)$, is determined by the function $h(t) = 29.4 + 24.5t - 4.9t^2$ where t is the number of seconds elapsed since the trajectory was thrown. What inequality best shows the time, in seconds, that the rock is in the air?

10. The weekly profits in Carlo's lawn care business can be modeled using the function $p(x) = -0.5x^2 + 40x + 300$ where profits are a function of the number of lawns Carlo and his crews can care for each week. Carlo's maximum profits earned are $1,100 for 40 lawns. In this situation what is the range of the function?

11. The number of bowling matches that must be scheduled in a league with n teams is given as $g(n) = \frac{1}{2}(n^2 - n)$ where each team plays every other team only one time. If the domain of the function for a league is the set of integers 1 to 10, what is the range of the function?

A all real numbers

B the set of integers 1 to 10

C {0, 1, 3, 6, 10, 15, 21, 28, 36, 45}

D the set of integers greater than or equal to 0 and less than or equal to 45

12. The function $y = (16 - x)(x)$ can be used to determine the area of a rectangular garden with a perimeter of 32 yards and a width of x yards. What is the possible range of the function that is reasonable for the situation?

F $0 \le y \le 4$

G $0 \le y \le 64$

H $0 \le x \le 4$

J $0 \le x \le 16$

WRITING QUADRATIC FUNCTIONS

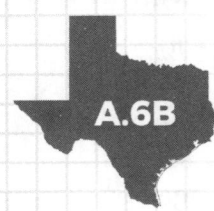

A.6B The student is expected to write equations of quadratic functions given the vertex and another point on the graph, write the equation in vertex form, $f(x) = a(x - h)^2 + k$, and rewrite the equation from vertex form to standard form, $f(x) = ax^2 + bx + c$.

ⓘ TELL ME MORE...

You can write quadratic equations in several ways. Two common ways to write quadratic equations are vertex form and standard form.

Vertex Form

$$f(x) = a(x - h)^2 + k$$

a, h, and k are real numbers. a represents the direction in which the parabola opens and a vertical dilation or compression. The point (h, k) represents the vertex of the parabola.

Standard Form

$$g(x) = ax^2 + bx + c$$

a, b, and c are real numbers. a represents the direction in which the parabola opens and a vertical dilation or compression. c represents the y-intercept. b provides information about horizontal transformations.

You can take an equation that is given to you in vertex form and use properties of algebra to rewrite it in standard form.

✓ EXAMPLES

EXAMPLE 1: The graph of $p(x)$ is shown. Write the function $p(x)$ in vertex form.

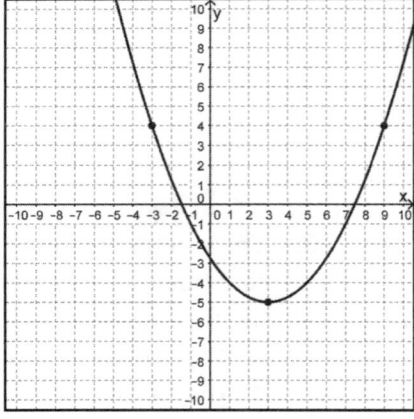

STEP 1 Determine the coordinates of the vertex of the parabola.

The vertex, (h, k), is the minimum point of this parabola and has the coordinates $(3, -5)$ in the graph.

Vertex: (3, –5)

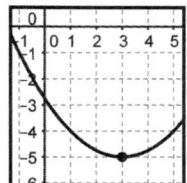

STEP 2 Determine the coordinates of an additional point on the graph of $p(x)$.

Locate $(-3, 4)$.

Second Point: (–3, 4)

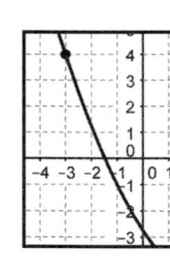

STEP 3 Vertex form of a parabola is $y = a(x - h)^2 + k$, where a is a multiplier and (h, k) is the vertex of the parabola. Substitute the coordinates of the vertex, $(3, -5)$, for h and k. Substitute the coordinates of the second point, $(-3, 4)$, for x and y.

Since $(h, k) = (3, -5)$, $h = 3$ and $k = -5$

Since $(x, y) = (-3, 4)$, $x = -3$ and $y = 4$

$4 = a(-3 - 3)^2 + (-5)$

STEP 4 Solve for a.

$$4 = a(-3 - 3)^2 + (-5)$$
$$4 = a(-6)^2 - 5$$
$$4 = 36a - 5$$
$$9 = 36a$$
$$\frac{1}{4} = a$$

$a = \frac{1}{4}$

STEP 5 Substitute a, h, and k into the vertex form of a quadratic function, $p(x) = a(x - h)^2 + k$.

$$p(x) = \frac{1}{4}(x - 3)^2 + (-5)$$

$p(x) = 1{-}4(x - 3)^2 - 5$

YOU TRY IT!

Write the equation of $m(x)$ in vertex form. The vertex of the graph of $m(x)$ is $(-7, 2)$ and the point $(-5, 14)$ lies on the graph of $m(x)$.

Substitute h, k, x, and y into vertex form, $y = a(x - h)^2 + k$ and solve for a.

$a =$ _____ $h =$ _____ $k =$ _____

Function: _____

EXAMPLE 2: Bernadette uses the function $c(x) = 0.5(x + 5)^2 + 2.3$ is used to predict the cost of manufacturing x baskets. However, a computer program that she uses to analyze profit and loss requires the function to be written in standard form, $y = ax^2 + bx + c$, in order to input the coefficients a, b, and c. Write $c(x)$ in standard form.

STEP 1 Use the properties of algebra to expand the binomial and simplify the expression. Rewrite $(x + 5)^2$ as $(x + 5)(x + 5)$ and multiply.

$c(x) = 0.5(x^2 + 10x + 25) + 2.3.$

$c(x) = 0.5(x + 5)^2 + 2.3$
$c(x) = 0.5(x + 5)(x + 5) + 2.3$
$c(x) = 0.5(x^2 + 5x + 5x + 25) + 2.3$
$c(x) = 0.5(x^2 + 10x + 25) + 2.3$

STEP 2 Distribute 0.5 (the value of a).

$c(x) = 0.5x^2 + 5x + 12.5 + 2.3$

$c(x) = 0.5(x^2 + 10x + 25) + 2.3$
$c(x) = 0.5x^2 + 5x + 12.5 + 2.3$

STEP 3 Combine like terms.

$c(x) = 0.5x^2 + 5x + 14.8$

$c(x) = 0.5x^2 + 5x + 12.5 + 2.3$
$c(x) = 0.5x^2 + 5x + 14.8$

1. The graph of $g(x)$, a quadratic function, has a vertex of $(-3, 6)$ passes through the point $(3, -12)$. Write $g(x)$ in standard form.

2. The graph of $f(x)$, a quadratic function, uses a scale factor of 4 and has a vertex of $(5, -3)$. Write $f(x)$ in standard form.

3. The function $f(x) = -6(x + 1)^2 - 7$ is plotted on a graph. The graph of the function $g(x)$ represents a transformation of $f(x)$ by shifting the vertex point 5 units right and 4 units up. Write $g(x)$ in both vertex and standard forms.

4. The graph of a quadratic function is shown below.

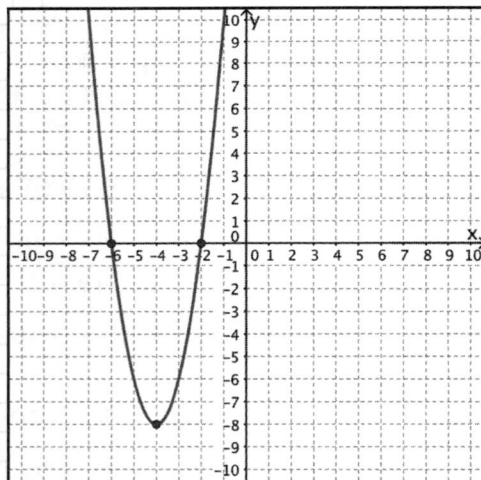

Write the function in vertex form.

5. The graph of $y = 64x - 16x^2$ models the height of a baseball in feet based on the time in seconds the ball is in the air.

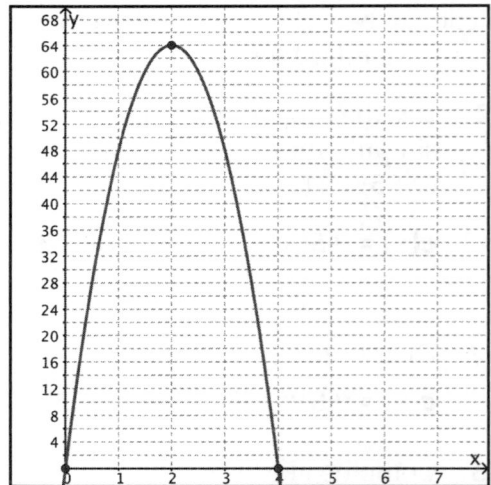

What are the values of a, h, and k that would be used to write the function in vertex form?

6. The graph of a company's production costs in dollars as related to the number of units produced is modeled by the function $p(x) = a(x - h)^2 + k$. What is the ordered pair that corresponds to the point of minimum cost? What is this point on the graph of the function?

8. The graph of a quadratic function has a maximum value of (2, 4) and includes the point (4, –8). Which of the following best describes this function?

A $y = -\frac{1}{72}(x - 4)^2 + 2$

B $y = -\frac{5}{32}(x - 2)^2 + 4$

C $y = -3(x - 2)^2 + 4$

D $y = -\frac{1}{3}(x + 2)^2 + 4$

7. The graph of the function $h(x)$ is shown.

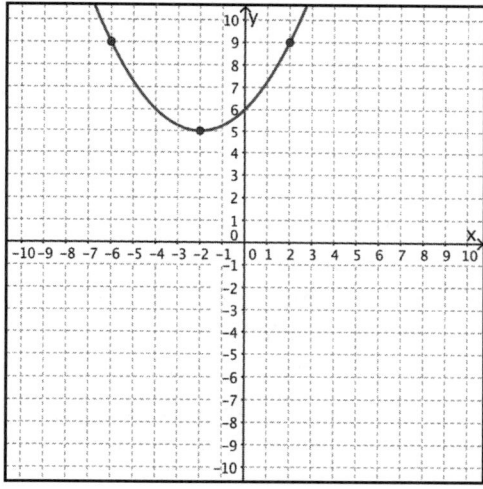

If $h(x)$ is written in vertex form, what is the value of a? Record your answer and fill in the bubbles on your answer document.

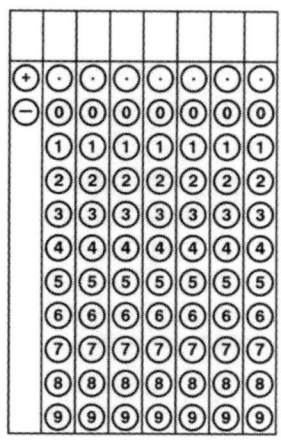

9. The function graphed below can be written in the form of $y = a(x - h)^2 + k$.

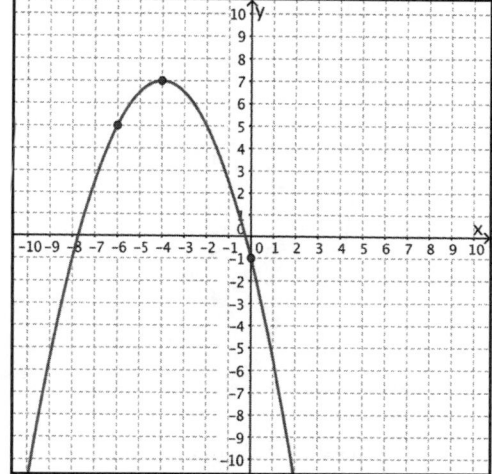

What are the values of a, h, and k for this function?

F $a = -\frac{1}{2}, h = 4, k = 7$

G $a = -\frac{1}{2}, h = 7, k = -4$

H $a = \frac{1}{2}, h = 7, k = 4$

J $a = -\frac{1}{2}, h = -4, k = 7$

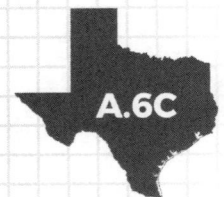

A.6C — The student is expected to write quadratic functions when given real solutions and graphs of their related equations.

(i) TELL ME MORE...

The zeros of a quadratic function, $f(x)$, are the solutions to the related equation $f(x) = 0$. If you know the solutions to this related equation, then you can use those solutions to determine the linear factors of $f(x)$ and then write possible symbolic forms of $f(x)$.

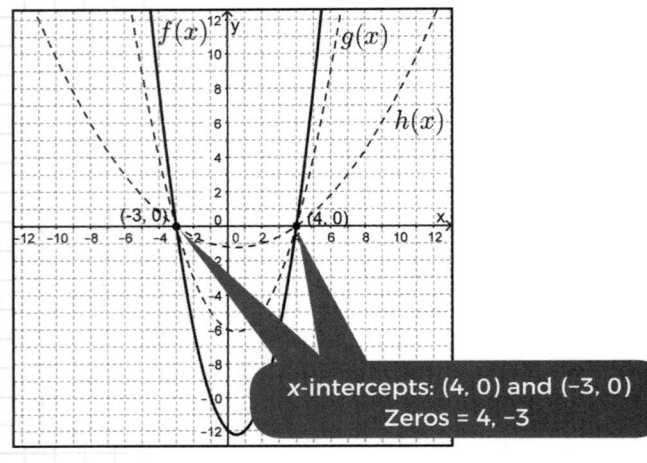

x-intercepts: (4, 0) and (–3, 0)
Zeros = 4, –3

Writing Quadratic Functions from Zeros

Solutions: $x = 4, -3$

Use the Zero Product Property to identify factors of $f(x)$.

$x = 4$, so $x - 4 = 0$ linear factor: $(x - 4)$

$x = -3$, so $x + 3 = 0$ linear factor: $(x + 3)$

Use the factors to write a possible $f(x)$.

$f(x) = (x - 4)(x + 3) = x^2 - x - 12$

If you can identify the zeros of the function from the graph of the function, then you can use those zeros to write the function symbolically. The key is that the x-coordinates of the x-intercepts are the same as the zeros of the quadratic function. It is also worth noting that many functions share the same zeros. In the graph shown, $f(x)$, $g(x)$, and $h(x)$ all have the zeros 4 and –3. The difference is the value of a, which induces a vertical dilation or reflection of the graph of the function. The coordinates of a third point are necessary to write the specific function $f(x)$.

✓ EXAMPLES

EXAMPLE 1: The solutions of an equation related to $k(x)$, a quadratic function, are $\frac{3}{2}$ and -7. Write the function $k(x)$ in factored form, $k(x) = a(cx - d)(mx - n)$, and standard form?

STEP 1   The solutions of the related equation are equivalent to the zeros of $k(x)$. Use the zeros to write linear factors of $k(x)$.

$x = \frac{3}{2}$ $x = -7$

$x - \frac{3}{2} = 0$ $x - (-7) = 0$

$2x - 3 = 0$ $x + 7 = 0$

The linear factors are 2x – 3 and x + 7.

STEP 2   Write $k(x)$ as a product of the parameter a and its linear factors.

k(x) = a(2x – 3)(x + 7)

STEP 3   Use the distributive property to write $k(x)$ in standard form.

$$k(x) = 2ax^2 + 11ax - 21a$$

$$k(x) = a(2x - 3)(x + 7)$$
$$k(x) = a(2x^2 + 14x - 3x - 21)$$
$$k(x) = a(2x^2 + 11x - 21)$$
$$k(x) = 2ax^2 + 11ax - 21a$$

EXAMPLE 2: The graph of $p(x)$ is shown. Write the function $p(x)$ in standard form, $p(x) = ax + bx + c$.

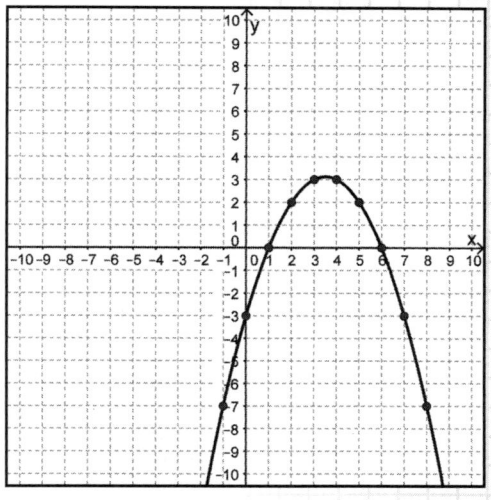

STEP 1   Determine the zeros of $p(x)$ from the x-intercepts in the graph.
The x-intercepts are $(1, 0)$ and $(6, 0)$, so the zeros of $p(x)$ are 1 and 6.

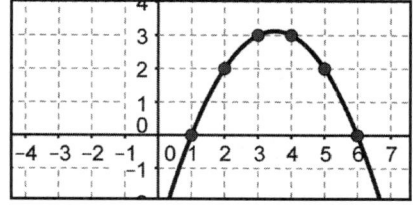

1 and 6

STEP 2   Use the zeros of $p(x)$ to write the linear factors of $p(x)$.

$x = 1$, so $x - 1 = 0$

$x = 6$, so $x - 6 = 0$

x – 1 and x – 6

STEP 3   Write $p(x)$ as a product of the parameter a and its linear factors.

p(x) = a(x – 1)(x – 6)

STEP 4   Select a third point, $(0, -3)$, on the graph of $p(x)$. Substitute its coordinates into the factored form of $p(x)$ to determine the value of a.

$$p(x) = a(x - 1)(x - 6)$$
$$-3 = a(0 - 1)(0 - 6)$$
$$-3 = a(-1)(-6)$$
$$-3 = 6a$$
$$-0.5 = a$$

a = –0.5

STEP 5   Use the distributive property and the value of a to write $p(x)$ in standard form.

$$p(x) = -0.5(x - 1)(x - 6)$$
$$p(x) = -0.5(x^2 - 6x - x + 6)$$
$$p(x) = -0.5(x^2 - 7x + 6)$$
$$p(x) = -0.5x^2 + 3.5x - 3$$

p(x) = –0.5x² + 3.5x – 3

YOU TRY IT!

Write the function $c(x)$ whose graph is shown.

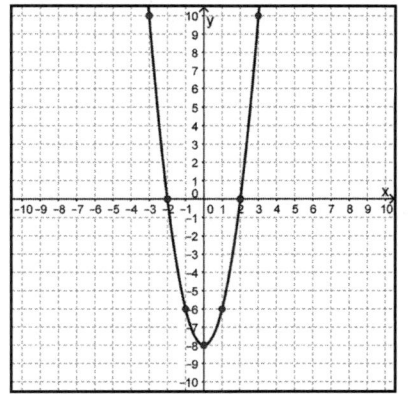

Zeros of $c(x)$: _____

Linear factors of $c(x)$: _____

Value of a: _____

Function: _____

For each graph shown, write the function in factored and standard form. Check your answer by using a graphing utility to see if the graph of your function matches the graph of the given function.

1.

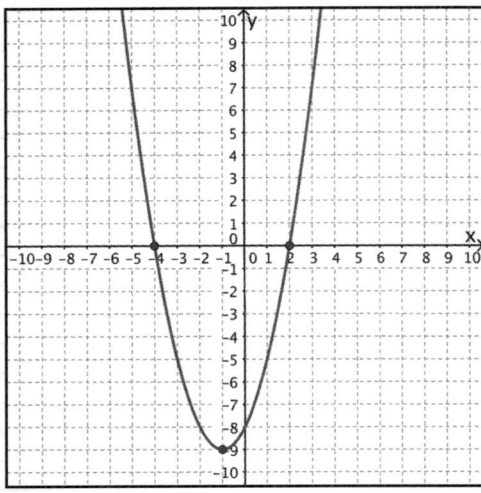

3.

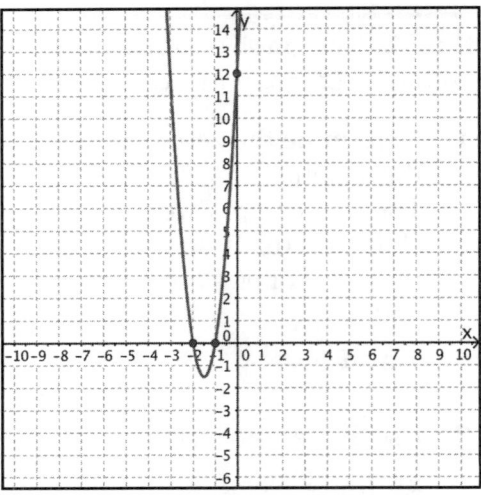

2.

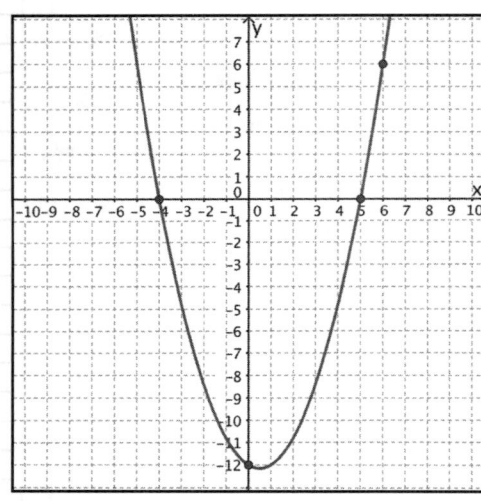

4.

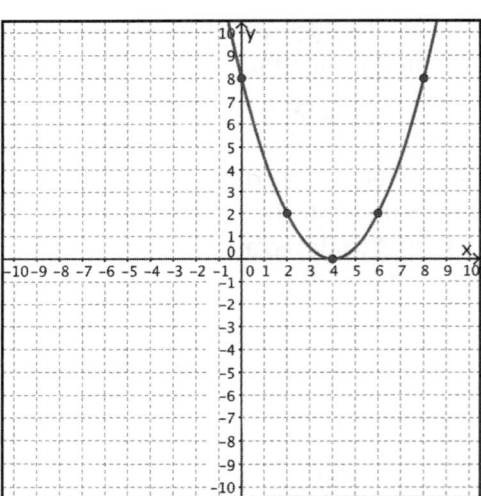

5. The graph of a quadratic function has *x*-intercepts at (–3, 0) and (1, 0). The graph also passes through the point (–2, –9). What is the function in standard form?

6. Tabitha has a business selling custom printed metal insulated cups. She uses the function graphed below, where x represents the number of cups to determine her profit, $p(x)$.

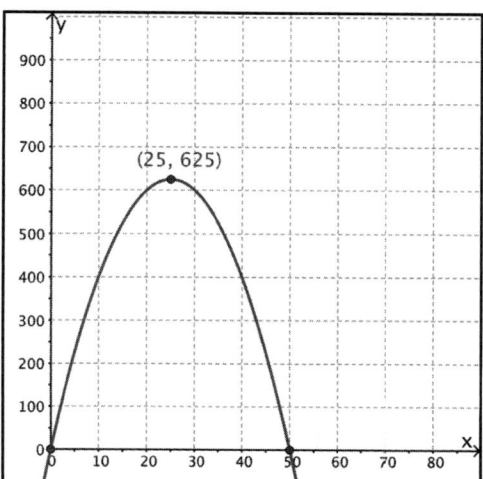

What function, in standard form, does Tabitha use?

7. The height, $h(x)$, of a golf ball in feet depends on the amount of time, x, in seconds the ball is in the air. The path of a golf ball is shown in the graph below. The zeros of the function equation are $x = 1$ and $x = 6$.

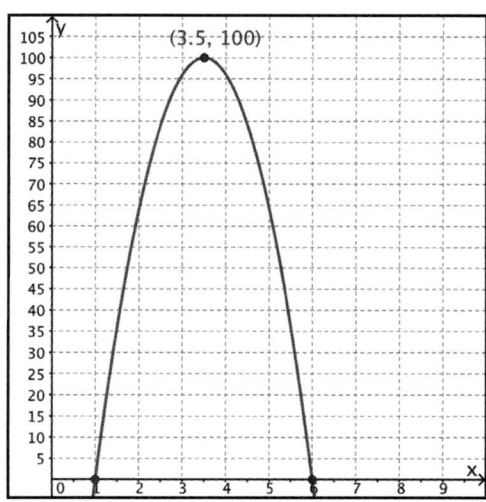

Write $h(x)$ in factored form.

8. The graph shows a parabola with zeros of -6 and $\frac{1}{9}$.

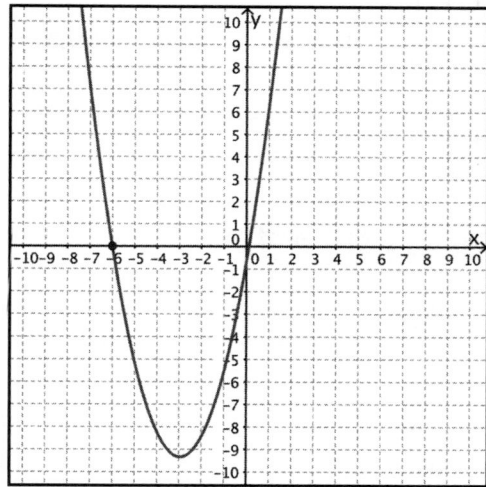

The point $(3, 26)$ lies on the graph of the parabola. What is the function whose graph is shown?

A $y = (x - 6)(x + \frac{1}{9})$

B $y = (x + 6)(x - \frac{1}{9})$

C $y = (x - 6)(x - \frac{1}{9})$

D $y = (x + 6)(x + \frac{1}{9})$

KEY ATTRIBUTES OF QUADRATIC FUNCTIONS

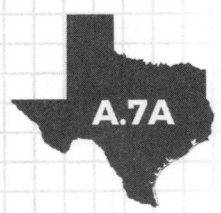

A.7A

The student is expected to graph quadratic functions on the coordinate plane and use the graph to identify key attributes, if possible, including x-intercept, y-intercept, zeros, maximum value, minimum values, vertex, and the equation of the axis of symmetry.

ⓘ TELL ME MORE...

The graph of a quadratic function reveals certain attributes that are important to the function.

For example, the graphs of $f(x) = -(x-4)^2 + 9$ and $g(x) = (x+5)^2 + 4$ are shown. Key features of the graphs include the x-intercept, y-intercept, zeros, maximum or minimum value, vertex, and axis of symmetry.

> Vertex of g: (−5, 4)
> If a > 0, the vertex is a minimum value.

> Vertex of f: (4, 9)
> If a < 0, the vertex is a maximum value.

> axis of symmetry: x = 4

> x-intercept: (1, 0)
> zero of f: 1

> x-intercept: (7, 0)
> zero of f: 7

> y-intercept: (0, -7)

- The **x-intercept** is the point(s) where the graph of the line crosses the x-axis (y = 0) A quadratic function may have 0, 1, or 2 x-intercepts.

- The **y-intercept** is the point where the graph of the line crosses the y-axis (x = 0).

- The **zero** of a quadratic function is the input value that generates an output value of 0. It is equivalent to the x-coordinate of the x-intercept. A quadratic function may have 0, 1, or 2 zeros.

- The graph of a quadratic function is a parabola. The **vertex** is a point that is a lower or upper bound for the parabola. If the parabola opens upward, like $g(x)$, the vertex is a lower bound for the parabola and it is a **minimum value**. If the parabola opens downward, like $f(x)$, the vertex is an upper bound for the parabola and it is a **maximum value**.

- The **axis of symmetry** is a vertical line of symmetry through the vertex.

✓ EXAMPLES

EXAMPLE 1: The graph of quadratic function h is shown. The coordinates of the x-intercepts, y-intercept and vertex are all integers. What are the vertex, minimum value, and axis of symmetry of the graph of h?

STEP 1 Identify the coordinates of the vertex from the graph.

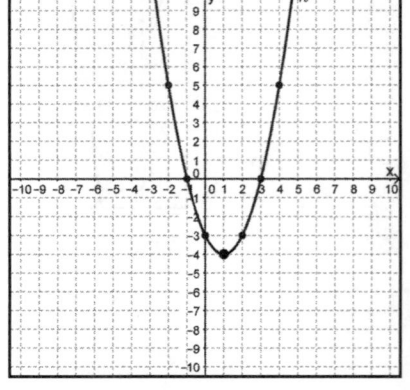

- The lower bound of the parabola is the point (1, −4).

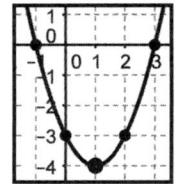

(1, −4)

STEP 2 Determine the minimum value using the y-coordinate of the vertex.

- The minimum value is the lowest function value on the parabola, which matches the y-coordinate of the vertex.

The minimum value is $y = -4$.

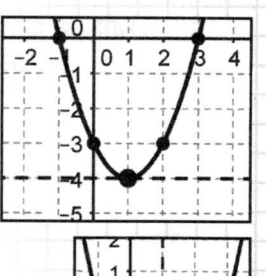

STEP 3 Determine the axis of symmetry using the x-coordinate of the vertex.

- The axis of symmetry is a vertical line through the vertex.

$x = 1$

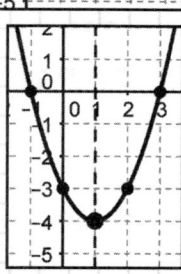

EXAMPLE 2: The graph of $p(x) = -(2x - 7)(x + 2)$ is shown. One zero of p is -2. What is the other zero of p? Record your answer and fill in the bubbles on your answer document.

STEP 1 Estimate the locations of the x-intercepts from the graph.

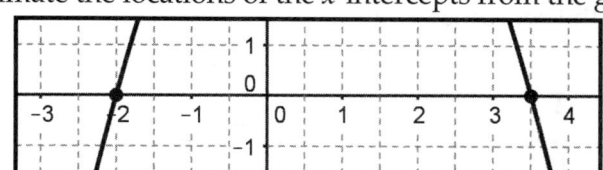

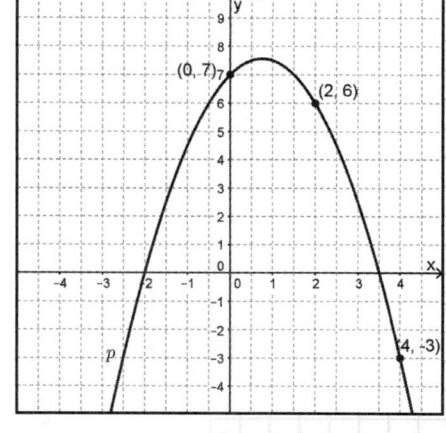

Near $(-2, 0)$ and $(3.5, 0)$

STEP 2 The zero of a function is the x-coordinate of an x-intercept. Verify the zeros algebraically from the function $p(x)$ using the x-coordinate of each x-intercept.

$$p(x) = -(2x - 7)(x + 2)$$
$$p(-2) = -(2(-2) - 7)((-2) + 2)$$
$$p(-2) = -(-4 - 7)(0)$$
$$p(-2) = -(-11)(0)$$
$$p(-2) = 0$$

$$p(x) = -(2x - 7)(x + 2)$$
$$p(3.5) = -(2(3.5) - 7)((3.5) + 2)$$
$$p(3.5) = -(7 - 7)(5.5)$$
$$p(3.5) = -(0)(5.5)$$
$$p(3.5) = 0$$

The second zero is 3.5.

STEP 3 Since the question is a gridded response question, enter your response on the grid provided. Practice using the grid with the instructions.

1. Record a 3 in the first column containing numbers. Record a decimal point in the next column. Record and a 5 in the next column.

2. Bubble the **3** beneath the numeral 3. Bubble the **.** beneath the decimal point. Bubble the **5** beneath the numeral 5.

EXAMPLE 3: The table shows several values for the quadratic function f. What are the coordinates of the x-intercept(s) and y-intercept?

x	-3	-2	-1	0	1	2	3
$f(x)$	-6	-7.5	-8	-7.5	-6	-3.5	0

STEP 1 Graph the points to see the structure of f.

STEP 2 Determine the y-intercept.

According to the table and graph, $f(0) = -7.5$.

(0, –7.5) is the y-intercept.

STEP 3 The graph shows two x-intercepts. Determine one x-intercept from the table.

According to the table, when $x = 3$, $f(x) = 0$.
Since $f(3) = 0$, this point represents an x-intercept.

(3, 0) is one x-intercept.

STEP 4 Estimate the other x-intercept from the graph. Use patterns in the table to confirm the coordinates of the second x-intercept.

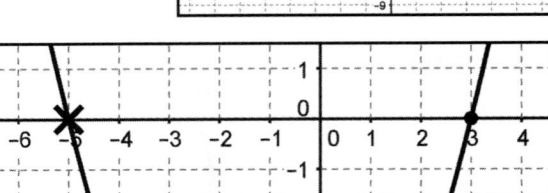

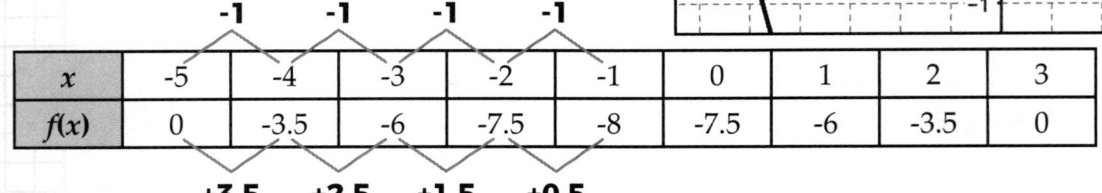

x	-5	-4	-3	-2	-1	0	1	2	3
$f(x)$	0	-3.5	-6	-7.5	-8	-7.5	-6	-3.5	0

(–5, 0) is the second x-intercept.

PRACTICE

For each graph shown, identify the vertex, maximum or minimum value, axis of symmetry, x-intercept, y-intercept, and zeros (if they exist).

1. $f(x) = 2(x - 3)^2 - 8$

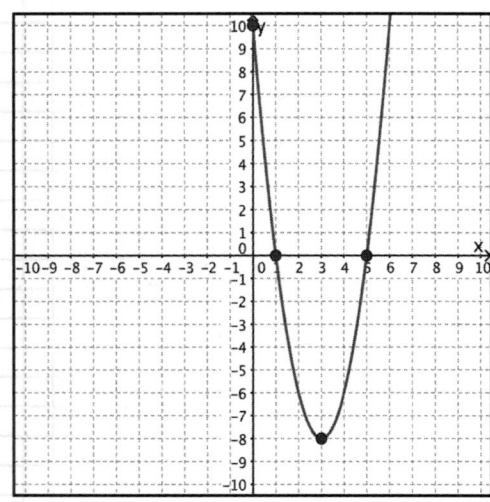

2. $g(x) = \frac{1}{2}(x - 1)^2 - 8$

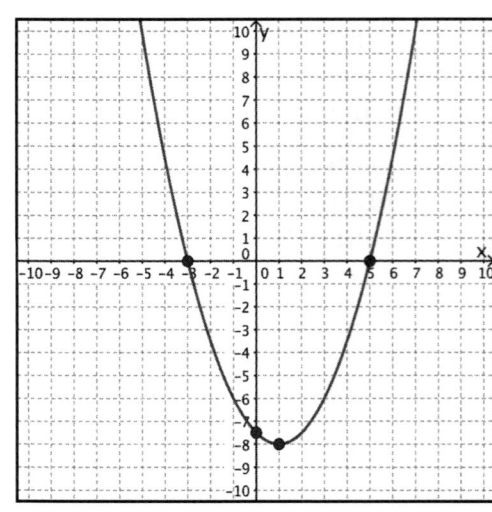

© Cosenza & Associates, LLC

3.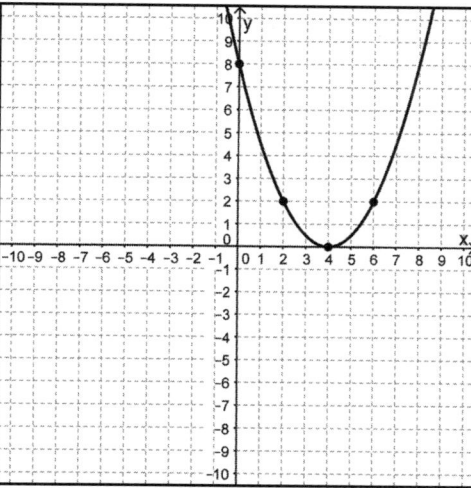

4.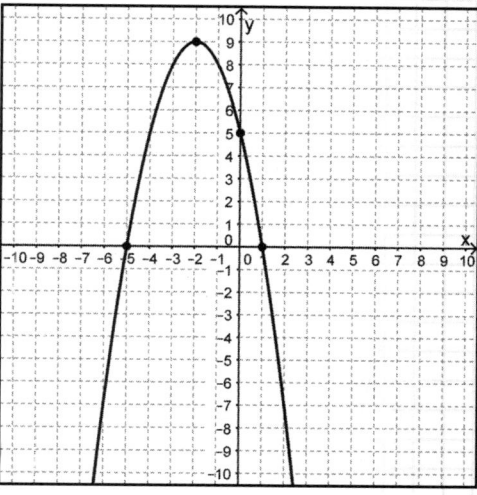

5. The function $d(t) = 19.6t - 4.9t^2$ represents the distance in meters an object is above the ground after t seconds when it is projected directly upward from the ground with an initial velocity of 19.6 meters per second. What is the maximum value of $d(t)$?

7. An archway in Mrs. Mayhill's office building measures 11 feet across at the base. The function modeling the archway is $a(x) = 11x - x^2$ where x is the distance in feet from one base of the archway and $a(x)$ represents the height of the arch. How tall is the arch at its vertex point and how far is this height from the base ends?

6. When the function $f(x) = (x + 2)^2 - 6$ is graphed, what is the equation of the axis of symmetry?

8. The graph of a quadratic function has a vertex point at $(-6, 12)$ and zeros of -8 and -4. What is the equation of the axis of symmetry for the graph of the function? Does the graph open upward or downward?

9. The graph of quadratic function k is shown below. What is the minimum value of k? Record your answer and fill in the bubbles on your answer document.

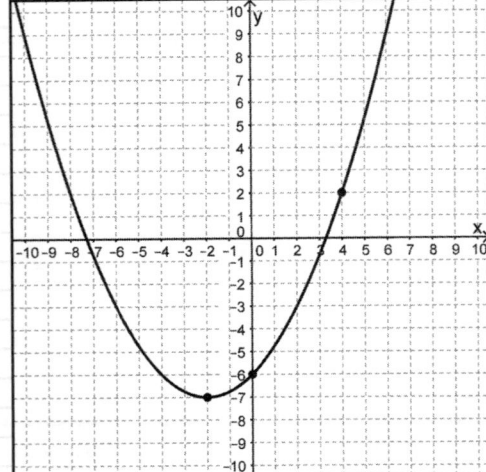

10. The graphs of quadratic functions p, q, and r are shown plotted below.

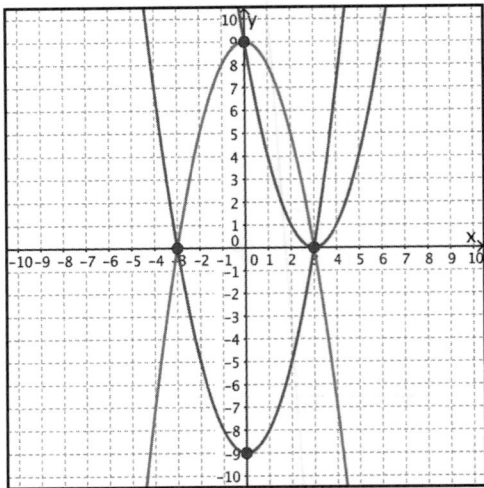

Which of the following of true of all three graphs?

A They all have the same axis of symmetry.

B They all have two x-intercepts.

C They all have the same y-intercept.

D They all share a zero at 3.

11. The table below shows values for a quadratic function.

x	-2	0	4	10
y	-24	-10	6	0

What are the values of the x-intercepts of the function?

F $(0, -10)$ and $(10, 0)$

G $(0, -10)$ and $(2, 0)$

H $(2, 0)$ and $(10, 0)$

J $(0, 2)$ and $(0, -10)$

LINEAR FACTORS AND ZEROS

A.7B The student is expected to describe the relationship between the linear factors of quadratic expressions and the zeros of their associated quadratic functions.

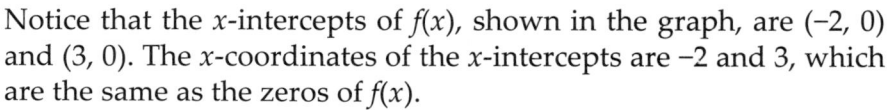
ℹ️ TELL ME MORE...

A **zero** of a polynomial function is a value of the independent variable (input value) that generates a function value (dependent variable or output value) of zero. If you were to graph the polynomial function, the zero would also be the x-coordinate of the x-intercept.

The quadratic expression $x^2 - x - 6$ factors to $(x + 2)(x - 3)$. Its associated quadratic function is $f(x) = (x + 2)(x - 3)$. The zeros of $f(x)$ can be identified using the Zero Product Property. Set each linear factor equal to 0 and then solve for x.

$$x + 2 = 0 \qquad\qquad x - 3 = 0$$
$$x = -2 \qquad\qquad x = 3$$

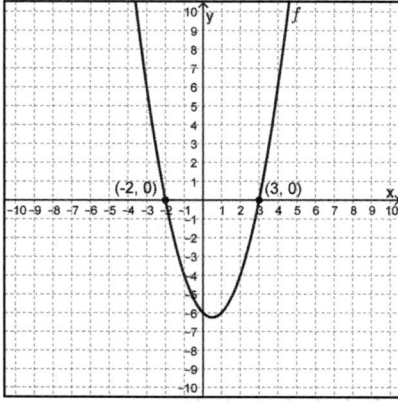

Notice that the x-intercepts of $f(x)$, shown in the graph, are $(-2, 0)$ and $(3, 0)$. The x-coordinates of the x-intercepts are -2 and 3, which are the same as the zeros of $f(x)$.

The quadratic expression $8x^2 + 3x - 4$ factors to $(2x - 1)(4x + 5)$. Its associated quadratic function is $g(x) = (2x - 1)(4x + 5)$. The zeros of $f(x)$ can be identified using the Zero Product Property. Set each linear factor equal to 0 and then solve for x.

$$2x - 1 = 0 \qquad\qquad 4x + 5 = 0$$
$$x = \frac{1}{2} \qquad\qquad x = -\frac{5}{4}$$

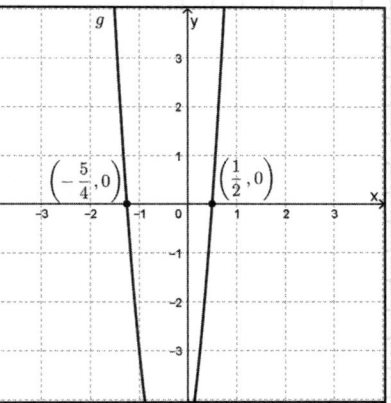

In general, if a quadratic expression factors to $n(ax - b)(cx - d)$, where a, b, c, d, and n are real numbers, the zeros of the associated function $f(x) = n(ax - b)(cx - d)$ are $\frac{b}{a}$ and $\frac{d}{c}$.

✓ EXAMPLES

EXAMPLE 1: What are the zeros of $k(x) = 3x^2 + 7x - 20$?

STEP 1 Factor completely the quadratic expression associated with $k(x)$, $3x^2 + 7x - 20$.

(3x – 5)(x + 4)

$$3x^2 + 7x - 20$$
$$3x^2 + 12x - 5x - 20$$
$$3x(x + 4) - 5(x + 4)$$
$$(3x - 5)(x + 4)$$

STEP 2   Set each linear factor equal to 0 and solve for x.

$3x - 5 = 0$ $x + 4 = 0$

$3x = 5$ $x = -4$

$x = \dfrac{5}{3}$

The zeros are $\dfrac{5}{3}$ and –4.

EXAMPLE 2: The table contains some values of $f(x) = -x^2 + 5x - 4$. What are the zeros of $f(x)$ and how do they relate to the linear factors of $f(x)$?

x	$f(x)$
-1	10
0	-4
1	0
2	2
3	2
4	0

STEP 1   Identify the zeros of $f(x)$ from the table. A zero is the x-value that generates $f(x) = 0$.

x	$f(x)$
-1	10
0	-4
1	0
2	2
3	2
4	0

The zeros are 1 and 4.

STEP 2   Factor completely $f(x) = -x^2 + 5x - 4$.

$$-x^2 + 5x - 4$$
$$-(x^2 - 5x + 4)$$
$$-(x - 4)(x - 1)$$

$-(x - 4)(x - 1)$

STEP 3   Compare the zeros of $f(x)$ to the linear factors of $f(x)$.

$$-(x - 4)(x - 1)$$

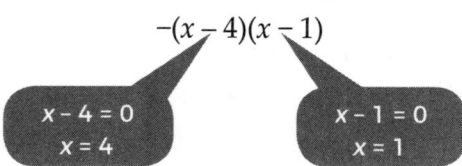

$x - 4 = 0$
$x = 4$

$x - 1 = 0$
$x = 1$

The zeros are 4 and 1, because $f(x) = -(x - 4)(x - 1)$.

YOU TRY IT!

Describe the relationship between the zeros of $g(x) = x^2 - 5x - 14$ and the linear factors of $g(x)$.

Linear Factors of g: _____ and _____

Zeros of g: _____ and _____

Relationship:

The zeros of g are _____ and _____ because $g(x)$ factors to _____.

EXAMPLE 3: The graph of quadratic function $r(x) = \frac{1}{2}x^2 - 3x + \frac{5}{2}$ is shown. Describe how the zeros of r are related to the linear factors of r.

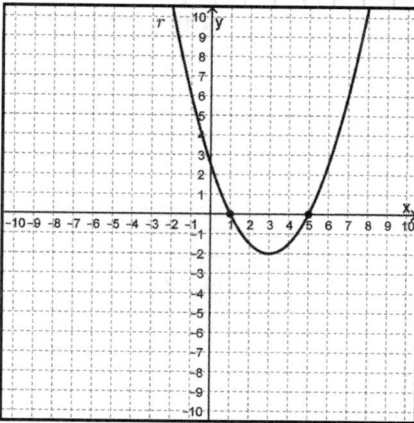

STEP 1 Determine the zeros of r from the graph. Locate the x-intercepts. The zeros are the x-coordinates of the x-intercepts.

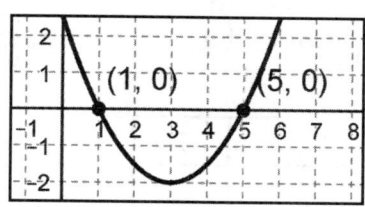

The zeros are 1 and 5.

STEP 2 Factor $r(x)$ so that it is written as a product of linear factors.

$$r(x) = \frac{1}{2}x^2 - 3x + \frac{5}{2}$$

$$r(x) = \frac{1}{2}(x^2 - 6x + 5)$$

$$r(x) = \frac{1}{2}(x - 1)(x - 5)$$

$$r(x) = \frac{1}{2}(x - 1)(x - 5)$$

STEP 3 Describe the relationship between the zeros from the graph and the linear factors of $r(x)$.

- The zeros are the x-coordinates of the x-intercepts of the graph of $r(x)$.
- The zeros are the x-values when each linear factor is set equal to 0.

The zeros of $r(x)$ are 1 and 5 because $r(x) = \frac{1}{2}(x - 1)(x - 5)$.

PRACTICE

1. What are the zeros of the quadratic function $f(x) = 2x^2 + 14x + 24$?

2. The zeros of a certain quadratic function $f(x)$ are p and q. The coefficient of the function is a. What is $f(x)$ when written in factored form?

3. Quadratic function $g(x) = 0.5x^2 - 2.5x + 2$ is graphed.

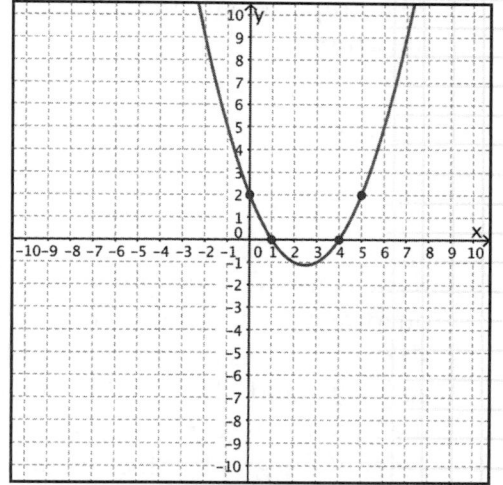

What is $g(x)$ written in factored form? What are the zeros of the function $g(x)$?

4. Some points that lie on quadratic function $h(x)$ are listed in the table. What is $h(x)$ written in factored form? What are the zeros of the function $h(x)$?

x	-2	0	3	7
y	21	5	-4	12

5. A golf pro uses the graph of the function $f(x) = -16x^2 + 80x$ to model the trajectory of a golf ball after he hits it in order to better understand his golf swing. When graphed what are the zeros of the graph that relate to the times where the ball is on the ground?

6. When $k(x) = -(x + 3)(x + 8)$ is plotted on a graph, what are the zeros of the graph?

7. Kylie constructs the graph of a quadratic function $k(x)$. The parabola she plots opens downward and has zeros of 3 and −1. The y-intercept of the function is (0, 6) and the vertex point of the graph is located at point (1, 8). What is $k(x)$ written in factored form?

8. The zeros of a quadratic function $p(x)$ are $\frac{3}{4}$ and $\frac{12}{5}$. The coefficient of the function is 1. What is $p(x)$ when written in factored form?

9. Which statement about $f(x) = x^2 - 12x + 36$ is true?

A The zero is 6, because $f(x) = (x - 6)^2$.

B The zeros are −6 and 6, because $f(x) = (x + 6)(x - 6)$.

C The zeros are 9 and 4, because $f(x) = (x - 9)(x - 4)$.

D The zero is −6, because $f(x) = (x + 6)^2$

10. Which statement is NOT true of the function $k(x) = 7 - 6x - x^2$?

F The graph opens downward.

G The function written in factored form is $k(x) = -(x + 7)(x - 1)$.

H The zeros of the function are 7 and −1.

J The vertex of the graph is (−3, 16).

TRANSFORMATIONS OF QUADRATIC FUNCTIONS

A.7C

The student is expected to determine the effects on the graph of the parent function $f(x) = x^2$ when $f(x)$ is replaced by $af(x)$, $f(x) + d$, $f(x - c)$, $f(bx)$ for specific values of a, b, c, and d.

ⓘ TELL ME MORE...

The quadratic parent function, $f(x) = x^2$, can be transformed by multiplying or adding real numbers to the independent variable, x, or the independent variable after it's squared, x^2. Different transformations of the parent function affect the graph of the parent function in different ways. The transformation parameter $b \neq 0$.

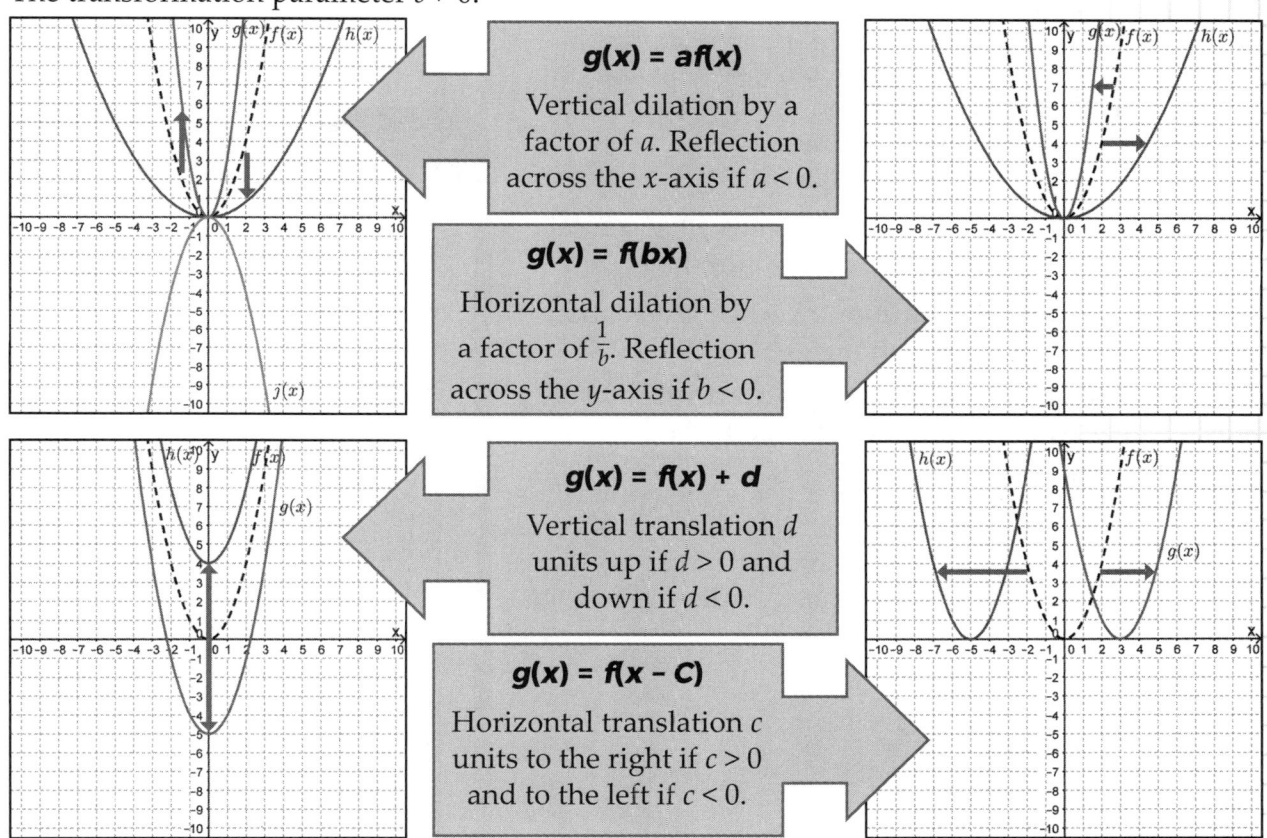

g(x) = af(x)

Vertical dilation by a factor of a. Reflection across the x-axis if $a < 0$.

g(x) = f(bx)

Horizontal dilation by a factor of $\frac{1}{b}$. Reflection across the y-axis if $b < 0$.

g(x) = f(x) + d

Vertical translation d units up if $d > 0$ and down if $d < 0$.

g(x) = f(x − C)

Horizontal translation c units to the right if $c > 0$ and to the left if $c < 0$.

✓ EXAMPLES

EXAMPLE 1: The graph of $f(x) = x^2$ was transformed to create the graph of $g(x) = 7.2x^2$. What transformation was done to f to generate g?

STEP 1   Identify the parameter that is added to or multiplied by f to create g.
$g = 7.2x^2$, so to create $g(x)$, you multiply 7.2 by f. This factor is the parameter a.

a = 7.2

STEP 2 Determine the effect of the parameter a on the graph of f. The parameter a causes a vertical dilation by a factor of a. Since $a = 7.2$, which is greater than 1, there will be a vertical stretch of the graph of f by a factor of 7.2 to create the graph of g. Since 7.2 is positive, there is no vertical reflection.

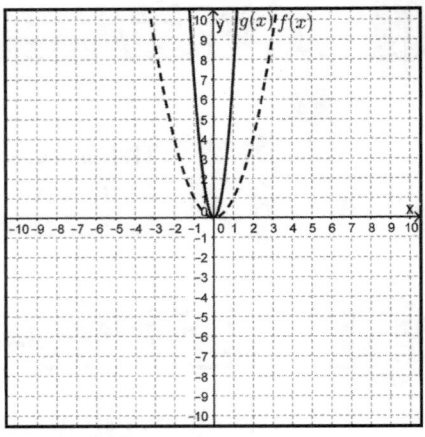

Vertically stretch f by a factor of 7.2.

STEP 3 Graph both f and g on the same screen or coordinate plane to confirm your answer.

The graph of f is vertically stretched by a factor of 7.2 to create the graph of g.

EXAMPLE 2: Arthur graphed quadratic functions p and q on the same coordinate grid. The vertex of the graph of q is 6 units to the right and 5 units up from the vertex of the graph of p. If $p(x) = x^2$, write the function for $q(x)$.

STEP 1 If the vertex is transformed, then all points on the parabola are transformed. Determine the transformations that occurred to the vertex of p to generate q and their associated parameters.

■ "6 units to the right" means a horizontal translation, which is the parameter, c.

■ "5 units up" means a vertical translation, which is the parameter, d.

The graph of q is translated horizontally (c) and translated vertically (d).

STEP 2 Since p was transformed using the parameters c and d, the function $q(x)$ is of the form $q(x) = (x - c)^2 + d$. Determine values for each parameter and use them to write the function for $q(x)$.

$$c = 6 \text{ and } d = 5$$

$$q(x) = (x - 6)^2 + 5$$

YOU TRY IT!

The graph of $f(x) = x^2$ was transformed to create the graph of $t(x) = (x + 1.5)^2$. What transformation was done to f to generate t?

Graph:

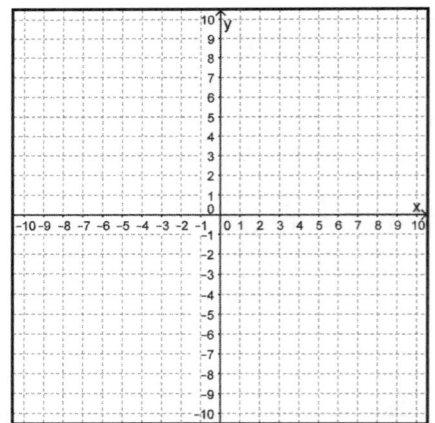

Does $t(x)$ include the parameter a, b, c, or d?

What transformation occurred to $f(x)$ to generate $t(x)$?

EXAMPLE 3: The graph of $f(x) = x^2$ is transformed to create the graph of $g(x)$ as shown. Write the function for $g(x)$ and describe the transformation.

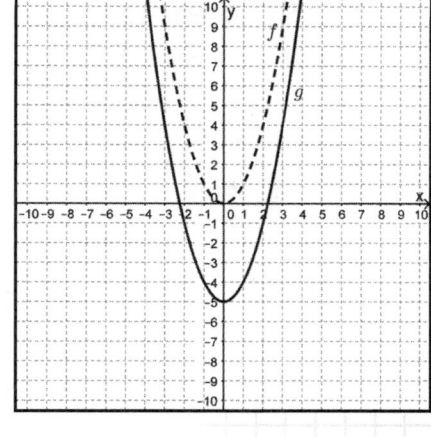

STEP 1 Visually inspect the graph to determine if a translation, reflection, or dilation has occurred. Determine if the transformation is vertical or horizontal and its magnitude.

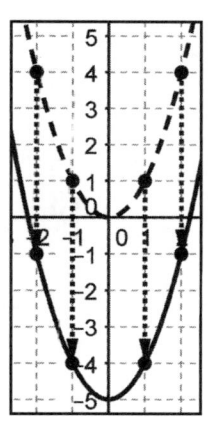

- The vertex of g is 5 units down from the vertex of f, suggesting a vertical translation has occurred.

- Inspecting the points near the vertex shows that all four points were also translated 5 units down.

f has been vertically translated 5 units down to create g.

STEP 2 Determine the parameter required to translate all points on the graph of f 5 units down to create the graph of g.

- A vertical translation is governed by the parameter d.

- Since the translation is down, the value of d is negative.

$d = -5$

STEP 3 Write the function for $g(x)$ using the value for d and the quadratic parent function, $f(x) = x^2$.

$g(x) = x^2 - 5$

PRACTICE

1. The graph of the quadratic parent function $f(x) = x^2$ is transformed to $h(x)$ by reflecting the graph across the x-axis, applying a vertical compression by a factor of $\frac{1}{2}$, a horizontal translation by 3 units to the right, and a vertical translation by 5 units down. What function represents $h(x)$?

2. The graph of the quadratic parent function, f(x), was transformed to create the graph of $h(x) = f(2x) - 4$. How was the parent function transformed to create h(x)?

3. The quadratic functions $f(x)$ and $g(x)$ are plotted on the same graph. The function $f(x)$ represents the quadratic parent function. The vertex of the graph of $g(x)$ is 6 units left and 8 units below the vertex of $f(x)$. What function represents $g(x)$?

4. The graph of $f(x)$ represents the quadratic parent function. The graph of $m(x)$ represents the function $m(x) = \frac{1}{4}f(x) + 4$. How does the graph of $m(x)$ compare to the graph of $f(x)$?

5. The graph of $f(x) = x^2$ is transformed to create the graph of $g(x) = af(x - c)^2 + d$ as shown. What function best represents $g(x)$?

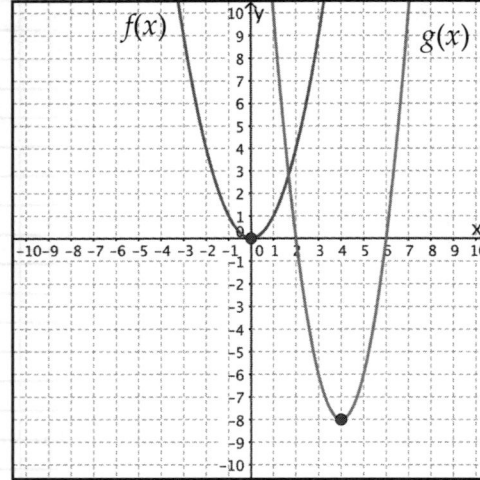

6. The graphs of functions A and B are shown on the graph below. Function B represents a transformation of Function A.

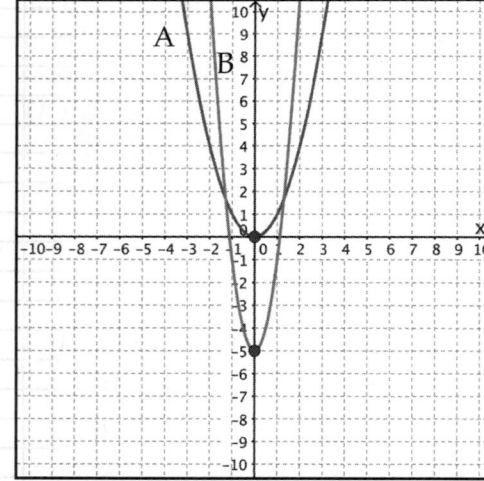

What transformation(s) was (were) applied to create Function B from Function A?

7. The table below represents points on the graph of a quadratic function $k(x)$ after the quadratic parent function, $f(x)$, was transformed to create $k(x)$.

x	$k(x)$
-2	1
-1	4
0	5
1	4
2	1

What transformations were applied to the graph of $f(x)$ to create $k(x)$?

8. Quadratic function $h(x)$ is plotted on a coordinate plane along with the quadratic parent function $f(x)$. The graph of $h(x)$ represents a transformation of $f(x)$ such that $h(x) = 2f(x + 1)^2$. Which of the following explains how the graph of $f(x)$ is changed to $h(x)$ with the transformations?

A The graph of $h(x)$ is wider and the vertex is shifted 1 unit left.

B The graph of $h(x)$ is wider and the vertex is shifted 1 unit up.

C The graph of $h(x)$ is narrower and the vertex is shifted 1 unit left.

D The graph of $h(x)$ is narrower and the vertex is shifted 1 unit down.

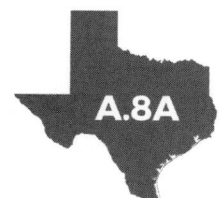

A.8A The student is expected to solve quadratic equations having real solutions by factoring, taking square roots, completing the square, and applying the quadratic formula.

ⓘ TELL ME MORE...

A **quadratic equation** is an equation that is based on a quadratic function. You can solve quadratic equations a variety of ways. You can use graphs or tables, as well as any number of algebraic methods. A quadratic equation is a second degree polynomial, so it may have 0, 1, or 2 real solutions.

Solving with Graphs

Treat the equation as though it were a system of equations.

- Define the function $Y1$ as the left member (side) of the equation and the function $Y2$ as the right member (side) of the equation.

- Graph $Y1$ and $Y2$ and determine the intersection point.

- The x-coordinate of the intersection point is the x-value that makes the function values equivalent, which means that both sides of the equation are equivalent and the equation is true.

Solving with Tables

Treat the equation as though it were a system of equations.

- Define the function $Y1$ as the left member (side) of the equation and the function $Y2$ as the right member (side) of the equation.

- Make a table of values for $Y1$ and $Y2$.

- Determine the row that has equivalent function values for both $Y1$ and $Y2$.

- The x-value in this row is the x-value that makes the function values equivalent, which means that both sides of the equation are equivalent and the equation is true.

✓ EXAMPLES

EXAMPLE 1: The function $L = 0.8T^2$ models the relationship between the length, L, of a pendulum in feet and the period, T, of the pendulum in seconds. What is the approximate period of a pendulum that has a length of 12 feet? Round to the nearest hundredth if necessary.

STEP 1 Write a quadratic equation related to the function. If the period of the pendulum is 12 feet, then $L = 12$, which is the output value of the function.

$$12 = 0.8T^2$$

STEP 2 Graph the left member of the equation as $Y1$ and the right member as $Y2$. Use technology as appropriate.

- $Y1 = 12$

- $Y2 = 0.8T^2$

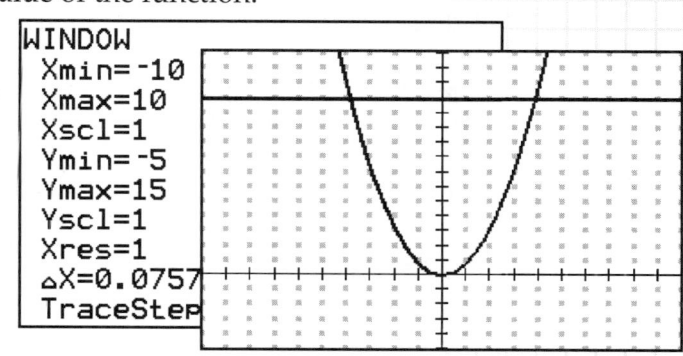

```
WINDOW
 Xmin=-10
 Xmax=10
 Xscl=1
 Ymin=-5
 Ymax=15
 Yscl=1
 Xres=1
 ▵X=0.0757
 TraceStep
```

STEP 3 Solve the system of equations, $Y1$ and $Y2$. Calculate the coordinates of the intersection point(s) of $Y1$ and $Y2$. The x-coordinate of each point represents a solution to the equation $12 = 0.8T^2$ since that is an x-value that makes both sides of the equation equivalent.

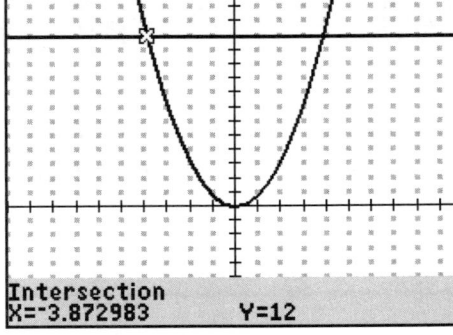
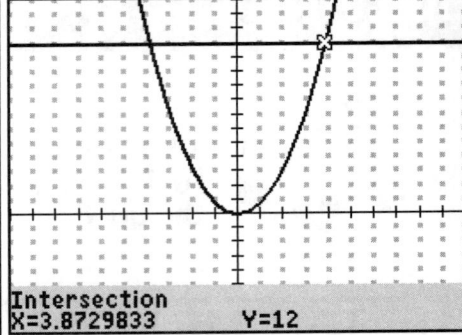

Intersection
X=-3.872983   Y=12

Intersection
X=3.8729833   Y=12

$x \approx -3.87, 3.87$

STEP 4 The question asked for the approximate period of the pendulum. In this context, the period is measured in time and cannot be negative, so the correct solution to the problem is the positive x-value that solves the equation.

$x \approx 3.87$

EXAMPLE 2: What are the solutions to the equation $8x^2 - 36x = 20$?

STEP 1 Make a table of values for x, the left member of the equation as $Y1$, and the right member as $Y2$. Use technology as appropriate.

■ $Y1 = 8x^2 - 36x$

■ $Y2 = 20$

x	-1	-0.5	0	0.5	1	1.5	2	2.5	3	3.5	4	4.5	5	5.5
$Y1$	44	20	0	-16	-28	-36	-40	-40	-36	-28	-16	0	20	44
$Y2$	20	20	20	20	20	20	20	20	20	20	20	20	20	20

STEP 2 Identify the column(s) containing equivalent values of $Y1$ and $Y2$. The x-value(s) in these column(s) represent solution(s) to the equation.

x	-1	-0.5	0	0.5	1	1.5	2	2.5	3	3.5	4	4.5	5	5.5
$Y1$	44	20	0	-16	-28	-36	-40	-40	-36	-28	-16	0	20	44
$Y2$	20	20	20	20	20	20	20	20	20	20	20	20	20	20

$x = -0.5, 5$

EXAMPLE 3: The area of a rectangular park with a perimeter of 200 yards can be modeled with the function $A(x) = 100x - x^2$, where x represents the length of the park. What is the largest length of a park with an area of 2100 square yards.

STEP 1 Write a quadratic equation related to the function. If the area is 2100 square yards, then $A(x) = 2100$, which is the output value of the function

STEP 2 Graph the left member of the equation as $Y1$ and the right member as $Y2$. Use technology as appropriate.

- $Y1 = 2100$
- $Y2 = 100x - x^2$

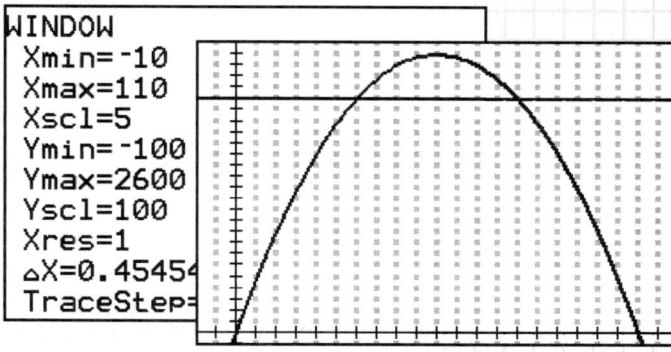

```
WINDOW
 Xmin=-10
 Xmax=110
 Xscl=5
 Ymin=-100
 Ymax=2600
 Yscl=100
 Xres=1
 ΔX=0.45454
 TraceStep=
```

STEP 3 Solve the system of equations, $Y1$ and $Y2$. Calculate the coordinates of the intersection point(s) of $Y1$ and $Y2$. The x-coordinate of each point represents a solution to the equation $2100 = 100x - x^2$ since that is an x-value that makes both sides of the equation equivalent.

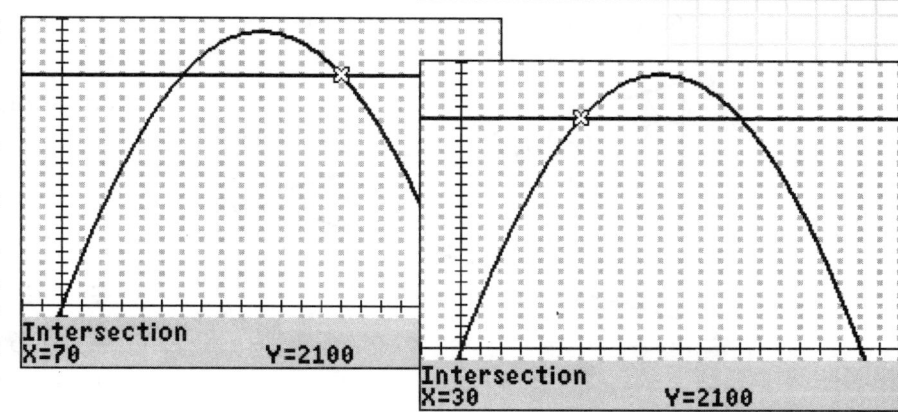

Intersection
X=70   Y=2100

Intersection
X=30   Y=2100

$x = 30, 70$

STEP 4 The question asked for largest length and 70 is greater than 30. Since the question is a gridded response question, enter your response on the grid provided. Practice using the grid with the instructions.

1. Record a **7** in the first column containing numbers and **0** in the next column.

2. Bubble the **7** beneath the numeral 7. Bubble the **0** beneath the numeral 0.

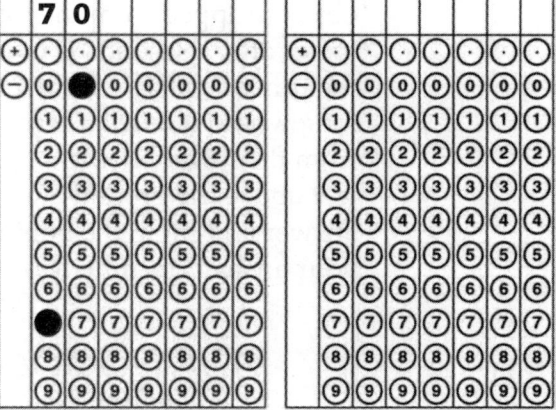

PRACTICE

Solve for x.

1. $7x^2 - 14x = -7$

2. $x^2 + 5x - 35 = 3x$

3. $6x^2 - 13x + 3 = -3$

4. The area of a rectangular garden is represented as $f(x) = 2x^2 - 3x + 30$ where x represents the length of the garden. What is the length of the garden if the area is 35 square yards?

7. Which of the following is closest to a solution of $m^2 - 12m + 26 = 0$?

   A 2.167

   B 2.837

   C 4.694

   D Not here

   $a = 1$
   $b = -12$
   $c = 26$

   $x^2 - 12x + 26 = 0$

   $x = \dfrac{12 \pm \sqrt{144 - 104}}{2}$

   $x = \dfrac{12 \pm \sqrt{40}}{2} = \dfrac{12 \pm 6.32}{2} = \dfrac{5.68}{2}$

   $= 2.84$

5. In the diagram, $\triangle HRC$ is an isosceles triangle. What are the values of x?

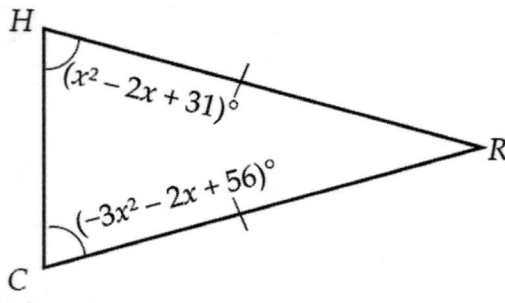

8. What is the solution set for the equation $5n^2 + 19n - 68 = -2$?

   F $\{2.2, -6\}$

   G $\{-2.2, 6\}$

   H $\{2, 6.6\}$

   J $\{-2, -6.6\}$

6. A fire watch tower in a state forest is 250 feet tall. The height, h, of an object dropped from the top of the tower after t seconds is modeled by the function $h(t) = -16t^2 + 250$. How long would it take an object dropped from the tower to reach a platform 54 feet above the ground? Record your answer and fill in the bubbles on your answer document.

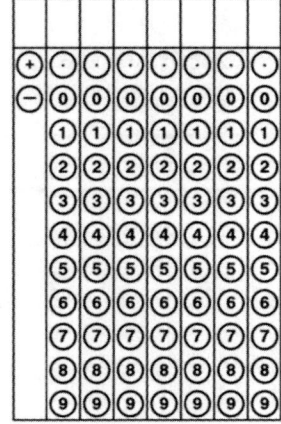

SOLVING QUADRATIC EQUATIONS: ALGEBRAIC METHODS

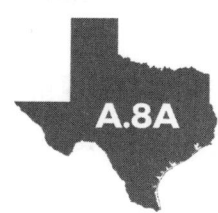

A.8A The student is expected to solve quadratic equations having real solutions by factoring, taking square roots, completing the square, and applying the quadratic formula.

ⓘ TELL ME MORE...

A **quadratic equation** is an equation that is based on a quadratic function. You can solve quadratic equations a variety of ways. You can use graphs or tables, as well as any number of algebraic methods. A quadratic equation is a second degree polynomial, so it may have 0, 1, or 2 real solutions.

Solving by Factoring The quadratic equation should be in standard form, $ax^2 + bx + c = 0$. ■ Completely factor the equation so that it is a series of factors with a product equal to 0. $$a(bx - c)(dx - e) = 0$$ ■ Apply the **Zero Product Property**. Set each factor equal to 0 and solve for x.	**Solving by Taking Square Roots** The quadratic equation should be in vertex form, $a(bx - c)^2 + d = 0$ or $a(bx - c)^2 = d$. ■ Apply inverse operations to solve for x. $$(bx - c)^2 = \frac{-d}{a}$$ ■ Take the square root of both sides. $$bx - c = \pm\sqrt{\frac{-d}{a}}$$ ■ Continue with inverse operations.
Solving by Completing the Square The goal is to write the quadratic equation in vertex form. For an equation in standard form, $ax^2 + bx + c = 0$: ■ Factor a from the x^2 and x terms. $$a\left(x^2 + \frac{b}{a}x\right) + c = 0$$ ■ Square half of $\frac{b}{a}$ and add it inside the parentheses. To balance this, subtract a times this quantity outside the parentheses. $$a\left(x^2 + \frac{b}{a}x + \left(\frac{b}{2a}\right)^2\right) + c - a\left(\frac{b}{2a}\right)^2 = 0$$ ■ Factor the expression inside the parentheses as a binomial squared. $$a\left(x + \frac{b}{2a}\right)^2 + c - a\left(\frac{b}{2a}\right)^2 = 0$$ ■ Continue with inverse operations.	**Solving with the Quadratic Formula** The quadratic equation should be in standard form, $ax^2 + bx + c = 0$. ■ Identify the values of a, b, and c as real numbers from the equation. ■ Substitute them into the quadratic formula: $$x = \frac{-b \pm \sqrt{b^2 - 4ac}}{2a}$$ ■ Simplify the radical expression.

✓ EXAMPLES

EXAMPLE 1: What are the solutions to $4(x - 5)^2 = 72$?

STEP 1 This equation has a binomial squared, like vertex form. Solve by taking square roots (inverse operations). First, take the inverse of multiplication. Divide by the coefficient of the binomial so that the binomial squared is by itself.

$$\frac{4(x - 5)^2}{4} = \frac{72}{4}$$
$$(x - 5)^2 = 18$$

$(x - 5)^2 = 18$

STEP 2 Take the inverse of the exponent. Take the square root of both sides of the equation. On the right side, be sure to include the positive or negative symbol, ±.

$x - 5 = \pm\sqrt{18}$

$$\sqrt{(x-5)^2} = \pm\sqrt{18}$$
$$x - 5 = \pm\sqrt{18}$$

STEP 3 Take the inverse of subtraction. Add 5 to both sides.

$x = 5 \pm \sqrt{18}$

$$x - 5 + 5\sqrt{18} + 5$$
$$x = 5 \pm \sqrt{18}$$

STEP 4 Simplify $\sqrt{18}$ by factoring 18 into 9 × 2.

$x = 5 \pm 3\sqrt{2}$

$$x = 5 \pm \sqrt{18}$$
$$x = 5 \pm \sqrt{9 \times 2} = 5 \pm (\sqrt{9})(\sqrt{2})$$
$$x = 5 \pm 3\sqrt{2}$$

EXAMPLE 2: The function $4x^2 - 20x$ describes the area of a series of swimming pools. If a swimming pool has an area of 140 square units, what is the value of x?

STEP 1 Write the quadratic equation showing the area for this particular swimming pool. Set the function equal to the known output value, $A = 140$ square units.

$4x^2 - 20x = 140$

STEP 2 Solve this equation by completing the square. Divide the entire equation by the common factor of 4 in preparation.

$$\left(\frac{4x^2}{4}\right) - \left(\frac{20x}{4}\right) = \frac{140}{4}$$

$x^2 - 5x = 35$

STEP 3 Since $a = 1$, divide b (which is −5) by 2 and then square the quotient. Add that quantity to both sides of the equation, including the factor of 4 from the left.

$$x^2 - 5x + (-2.5)^2 = 35 + (-2.5)^2$$
$$x^2 - 5x + 6.25 = 35 + 6.25$$
$$(x - 2.5)^2 = 41.25$$

$(x - 2.5)^2 = 165$

STEP 4 Use inverse operations to solve for x.

$$(x - 2.5)^2 = 41.25$$
$$\sqrt{(x-2.5)^2} = \pm\sqrt{41.25}$$
$$x - 2.5 = \pm\sqrt{41.25}$$
$$x - 2.5 + 2.5 = 2.5 \pm \sqrt{41.25}$$
$$x = 2.5 \pm \sqrt{41.25}$$

$x = 2.5 \pm \sqrt{41.25}$

YOU TRY IT!

Solve for x: $3x^2 + 11x = 4$.

Rewrite in standard form, equal to 0.

_____ = 0

Factor the polynomial completely.

_____ = 0

Set each factor equal to 0 and solve for x.

_____ = 0 _____ = 0

Solutions:_____

EXAMPLE 3: What is the positive solution to $2x^2 + 3x - 35 = 0$? Record your answer and fill in the bubbles on your answer document.

STEP 1   Look to see if the polynomial factors. $2(-35) = -70$, so look for factors of -70 that add up to 3.

$$-1 + 70 = 69 \qquad -2 + 35 = 33 \qquad -5 + 14 = 9 \qquad \boxed{-7 + 10 = 3}$$

Use –7 and 10

STEP 2   Factor $2x^2 + 3x - 35 = 0$.

(2x – 7)(x + 5) = 0

$$2x^2 + 3x - 35 = 0$$
$$(2x - 7)(x + 5) = 0$$
Check by multiplying:
$$2x^2 + 10x - 7x - 35 = 0$$
$$2x^2 + 3x - 35 = 0 \checkmark$$

STEP 3   Apply the Zero Product Property. Set each linear factor equal to 0 and solve for x.

$$\begin{array}{ll} 2x - 7 = 0 & x + 5 = 0 \\ 2x = 7 & x = -5 \\ x = 3.5 & \end{array}$$

x = –5, 3.5

STEP 4   The question asked for the positive solution, which is 3.5 ($-5 < 0$ and is not positive). Since the question is a gridded response question, enter your response on the grid provided. Practice using the grid with the instructions.

1. Record a **3** in the first column containing numbers and a decimal in the next column. Record a **5** in the third column.

2. Bubble the **3** beneath the numeral 3. Bubble the **.** beneath the decimal point. Bubble the **5** beneath the numeral 5.

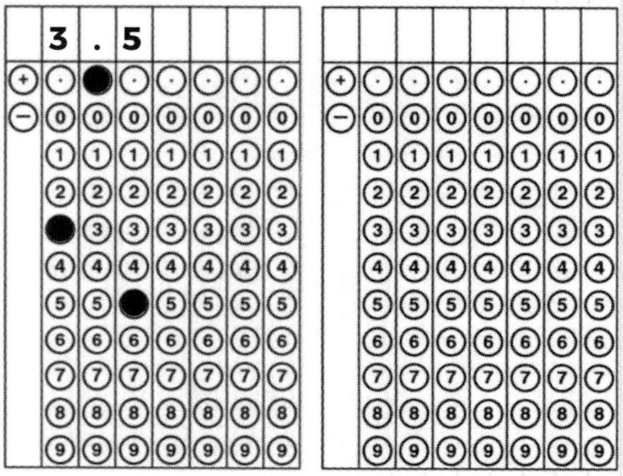

EXAMPLE 4: In the diagram, $\angle JAK$ is adjacent to $\angle KAL$. $m\angle JAK = (5x^2 - 8x)°$ and $m\angle KAL = 40°$. If $m\angle JAL = 95°$, what is the value of x?

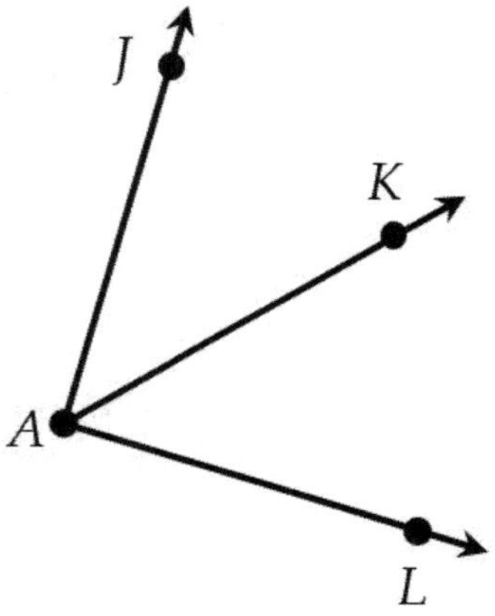

STEP 1   Use the angle addition postulate to write an equation in terms of x.

$$m\angle JAK + m\angle KAL = m\angle JAL$$
$$(5x^2 - 8x)° + (40°) = 95°$$

5x² – 8x + 40 = 95

STEP 2   Subtract 95 from both sides.

$$5x^2 - 8x + 40 = 95$$
$$5x^2 - 8x - 55 = 0$$

5x² – 8x – 55 = 0

STEP 3   Since the resulting trinomial doesn't factor, use the quadratic formula, $x = \dfrac{-b \pm \sqrt{b^2 - 4ac}}{2a}$. Substitute $a = 5$, $b = -8$, and $c = -55$.

$$x = \frac{-(-8) \pm \sqrt{(-8)^2 - 4(5)(-55)}}{2(5)}$$

$x = \dfrac{-(-8) \pm \sqrt{(-8)^2 - 4(5)(-55)}}{2(5)}$

STEP 4 Simplify the expression, including the radicand, or expression beneath the radical symbol.

$$x = \frac{4}{5} \pm \frac{\sqrt{291}}{5}$$

$$x = \frac{-(-8) \pm \sqrt{(-8)^2 - 4(5)(-55)}}{10}$$

$$x = \frac{8 \pm \sqrt{64 + 1100}}{10}$$

$$x = \frac{8 \pm \sqrt{1164}}{10}$$

$$x = \frac{8 \pm \sqrt{4 \cdot 291}}{10}$$

$$x = \frac{8 \pm 2\sqrt{291}}{10}$$

$$x = \frac{8}{10} \pm \frac{2\sqrt{291}}{10}$$

$$x = \frac{4}{5} \pm \frac{\sqrt{291}}{5}$$

PRACTICE

1. $x^2 - 8x + 5 = 0$

2. $\frac{1}{2}x^2 - 32 = 0$

3. $6x^2 + 11x = 35$

4. $6x^2 - 23x + 15 = 0$

5. The area of a rectangular garden is represented as $f(x) = x^2 + 6x + 8$ where x represents the length of the garden. What are the length and width measurements if the area of the garden is 80 square yards?

6. The sum of the squares of two consecutive odd integers is 202 and can be found using $n^2 + (n + 2)^2 = 202$ where $n \geq 1$. What is the value of the larger of the integers?

7. A military helicopter dropped food supplies to a troop training in the wilderness from a height of 784 feet above the campsite using a small parachute. The approximate height, h, of the food supply parachute after t seconds is given by the function $h(t) = -16t^2 + 784$. How many seconds did it take the food supplies to hit the ground? Record your answer and fill in the bubbles on your answer document.

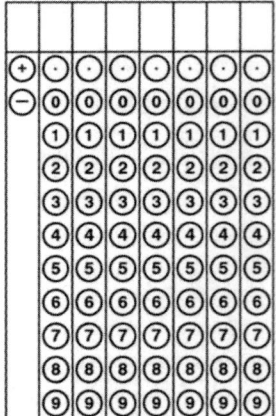

8. Some fireworks were projected vertically into the air from the ground at an initial velocity of 80 feet per second. The function modeling the path of the fireworks is given as $h(t) = -16t^2 + 80t$ where h represents the height of the fireworks t seconds after launch. How long where the fireworks in the air before returning to the ground?

9. John is 7 years older than his little brother Bobby. The product of John and Bobby's ages is 60. The situation can be represented as $b^2 + 7b = 60$ where b represents Bobby's age. How old are the two brothers?

10. Jamal solved the equation $x^2 + 6x + 4 = 0$ using the process of completing the square. His work is shown below.

Step 1: $x^2 + 6x - 4 = 0$

Step 2 $x^2 + 6x + 9 - 4 = 0$

Step 3: $(x^2 + 6x + 9) - 4 = 0$

Step 4: $(x + 3)^2 = 4$

Step 5: $x + 3 = \pm\sqrt{4}$

Step 6: $x = -3 \pm \sqrt{4}$

What mistake did Jamal make, if any, in his process?

11. Mr. Yu models his company's profits using the equation $y = 4x^2 - 12x$ where y is the amount of profit earned in thousands of dollars and x is the number of years in operation. Mr. Yu wants to project when his potential profits will be approximately 40 thousand dollars. How many years will Mr. Yu's company need to be in operation to make this profit?

A 2 years

B 5 years

C 20 years

D 10 years

12. What is the solution set to the equation $6n^2 - 18n - 18 = 6$?

F $\{-1, 4\}$

G $\{1, 2\}$

H $\{4, 27\}$

J $\{3, -3\}$

MODELING WITH QUADRATIC FUNCTIONS

A.8B

The student is expected to write, using technology, quadratic functions that provide a reasonable fit to data to estimate solutions and make predictions for real-world problems.

ⓘ TELL ME MORE...

Quadratic functions are second-degree polynomials functions. You can write a quadratic function in general form, $f(x) = ax^2 + bx + c$, where a, b, and c are real numbers.

$$f(x) = ax^2 + bx + c$$

Modeling with Transformations	**Modeling with Regression**
Make a scatterplot of the data. Beginning with the parent function, use transformations to move the vertex and then reflect and dilate the parabola.	Enter the data into a spreadsheet, graphing technology, or app. Use the technology's regression programming to generate a curve of best fit. Graph the curve on top of a scatterplot of your data to make sure the curve fits.

✓ EXAMPLES

EXAMPLE 1: A ride at an amusement park is 300 feet tall. At the top of the ride, a penny fell out of Tosha's pocket. The table shows the height, $h(x)$, of the penny x seconds after it fell. Write a function for $h(x)$ and use it to determine when the penny reached the ground.

Time (seconds), x	Height (feet), $h(x)$
0	300
1	284
2	236
3	156
4	44

STEP 1 Make a scatterplot of the data. Graph the quadratic parent function, $f(x) = x^2$, over the scatterplot.

STEP 2 Transform the vertex of the parent function, $f(x)$ so that it aligns with the maximum value, (0, 300).

■ Move the vertex up 300 feet to (0, 300) by adding 300 to the parent function. Or, let $d = 300$ in $f(x) = x^2 + d$.

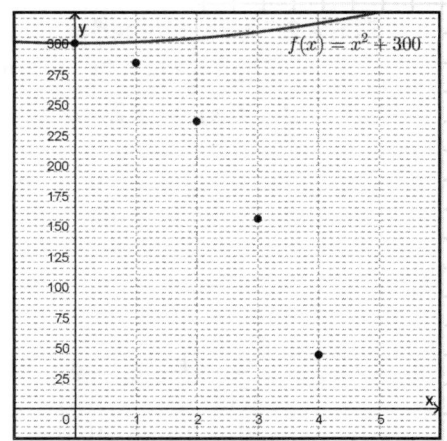

STEP 3 Reflect the graph vertically and dilate the graph so that it aligns with the data points.

■ Multiply $f(x)$ by −1, or let $a = −1$ in $f(x) = ax^2 + d$.

■ Multiply $f(x)$ by 16, or let $a = −16$ (factor of −1 from the reflection) in $f(x) = ax^2 + d$.

$h(x) = −16x^2 + 300$

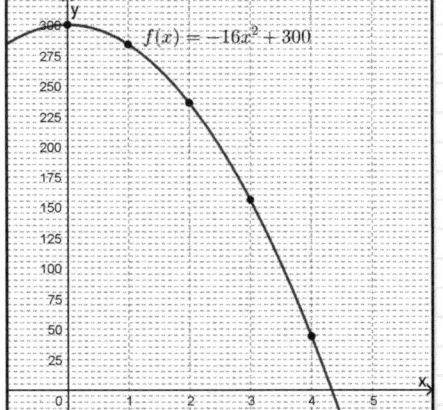

STEP 4 When the penny reaches the ground, $h(x) = 0$. Substitute $h(x) = 0$ as the function value and then solve for x.

$$0 = -16x^2 + 300$$

$$=300 = -16x^2$$

$$\frac{-300}{-16} = \frac{-16x^2}{-16}$$

$$18.75 = x^2$$

$$\pm\sqrt{18.75} = x$$

$$\pm 4.33 \approx x$$

The penny will reach the ground about 4.33 seconds after it begins to fall.

EXAMPLE 2: The average high temperature, $T(x)$, in degrees Fahrenheit, for each month, x, in El Paso, Texas, is shown in the table. Based on the table, write a quadratic function that best models the data.

STEP 1 Use regression with technology to determine an appropriate function model. Enter the data into your technology.

L1	L2	L3	L4	L5
1	58	------	------	------
2	63			
3	70			
4	79			
5	88			
6	96			
7	95			
8	92			
9	88			
10	78			
11	66			

L2(1)=**58**

Month, x	Temperature (°F), $T(x)$
1	58
2	63
3	70
4	79
5	88
6	96
7	95
8	92
9	88
10	78
11	66
12	57

STEP 2 Use the technology's quadratic regression algorithm to generate a quadratic regression equation.

$$T(x) = -1.2602x^2 + 16.8307x + 36.3636$$

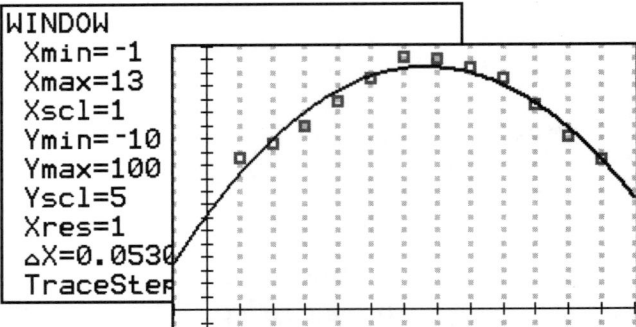

```
                    QuadReg
            y=ax²+bx+c
            a=-1.26023976
            b=16.83066933
            c=36.36363636
            R²=0.9418529739
```

STEP 3 Graph the quadratic regression equation for $T(x)$ over a scatterplot of the data. Compare the graph of the function model with the scatterplot of the actual data.

The graph does not pass through every data point in the scatterplot. However, it follows the general trend of the data.

```
WINDOW
 Xmin=-1
 Xmax=13
 Xscl=1
 Ymin=-10
 Ymax=100
 Yscl=5
 Xres=1
 △X=0.0530
 TraceSteⱼ
```

EXAMPLE 3: A pentagonal number is a number that can be represented by points arranged in a sequence of nested regular pentagons. The first five pentagonal numbers are shown in the diagram. If the pattern continues, write a function, $p(n)$, that can be used to determine the number of points in the nth pentagonal number. Then, use your function to predict the 15th pentagonal number.

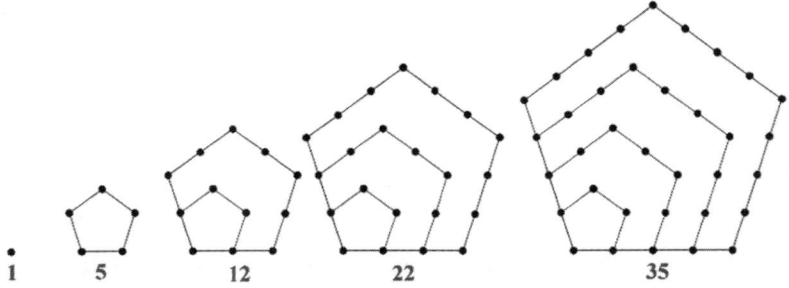

1   5   12   22   35

STEP 1 Make a table of data.

Term Number, n	1	2	3	4	5
Pentagonal Number, $p(n)$	1	5	12	22	35

STEP 2 Use regression with technology to determine an appropriate function model. Enter the data into your technology.

```
L1      L2      L3      L4      L5
1       1       ------  ------  ---
2       5
3       12
4       22
5       35
------  ------
```

STEP 3 Use the technology's quadratic regression algorithm to generate a quadratic regression equation.

p(n) = 1.5n² – 0.5n

```
QuadReg
y=ax²+bx+c
a=1.5
b=-0.5
c=0
R²=1
■
```

STEP 4 Determine the 15th pentagonal number. In this case, the term number (input value) is n, which is 15. Substitute $n = 15$ into $p(n)$ and evaluate the function.

The 15th pentagonal number is 330.

$p(n) = 1.5n^2 - 0.5n$

$p(15) = 1.5(15)^2 - 0.5(15)$

$p(15) = 1.5(225) - 0.5(15)$

$p(15) = 337.5 - 7.5$

$p(15) = 330$

PRACTICE

1. In the fall a local community organization hosts a pumpkin patch. As part of the pumpkin patch the group holds a "pumpkin chunkin'" contest where contestants try to catapult a pumpkin the farthest distance. The table below shows the distance traveled, in feet, of several pumpkins based on the angle setting of the catapult, in degrees.

Angle, x	Distance, y
20	360
30	462
40	509
50	501
60	437
70	318

Based on the data in the table, what function can best be used to model the situation?

2. The fuel efficiency of a certain vehicle, $f(x)$, is based on speed traveled, x. The table below shows the vehicle's miles per gallon at various speeds.

Miles per hour, x	Miles per gallon, y
20	14.2
24	17
30	20.3
40	23.4
44	23.8
50	23.5

Based on the table in the table, what is the approximate fuel efficiency of the vehicle when driven at an average rate of 70 miles per hour?

3. The following figures form a pattern.

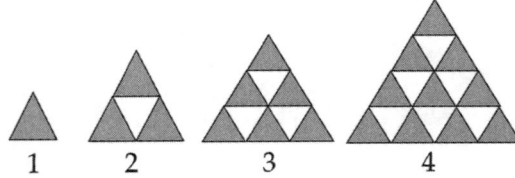

1 2 3 4

If the pattern continues, what expression can be used to find the number of shaded triangles that make up Figure n?

4. The graph below shows the amount of profit on soda sales for a concession stand at a small movie theatre based on the price of a soda.

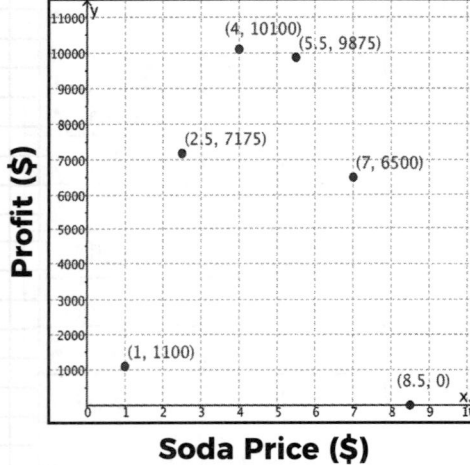

Soda Price ($)

Based on the data values in the graph, what function, $p(x)$, can best be used to model the situation?

For questions 5 and 6, use the following information.

The area of rectangle is modeled by the graph below where y is the area in square centimeters and x is the width in centimeters.

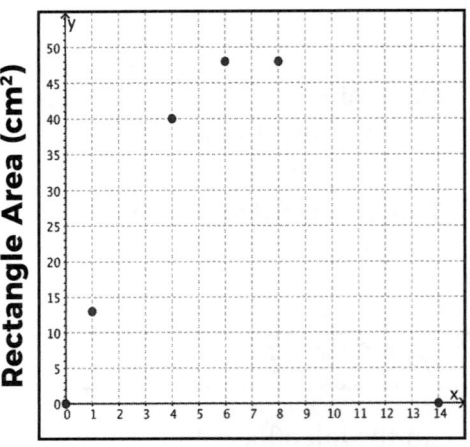

Rectangle Width (cm)

5. What function best represents the graph and could be used to determine any area, $f(x)$, given any possible width, x, for the rectangle.

6. What is the maximum area of the rectangle and what are the dimensions of the rectangle?

7. Cameron is practicing at the batting cages. He hits a baseball from a height of 5 feet off the ground. The table below shows the height of the baseball, $h(t)$, at different times.

Time (seconds), x	Height (feet), $h(t)$
0.5	26
1	39
1.5	44
2	41
2.5	30
3	11

Based on the table, what function can best be used to model the situation?

A $h(t) = -16t^2 + 5$

B $h(t) = -16t^2 + 50t + 5$

C $h(t) = 6t^2 + 5t + 22$

D $h(t) = 52t^2 + 5$

8. Malaki is making patterns with color tiles. The first 3 figures in his pattern are shown below. Each figure is made up of identical square tiles.

 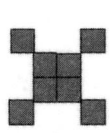 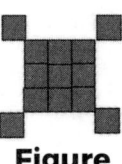

Figure 1   **Figure 2**   **Figure 3**

If the pattern continues, what expression can be used to find the number of square tiles that make up Figure n?

F $n^2 + n$

G $n^2 + 2n$

H $n^2 + 4$

J $4n^2$

9. The number of mosquitos in a city after a rainstorm can be modeled by quadratic function m. The table below shows the estimated number of mosquitos in millions, $m(x)$, present in the city based on the number of inches of rainfall, x.

Rainfall (inches), x	Mosquitos (millions), $m(t)$
0.5	6
1	11
1.5	16
2	20
3	27
3.5	30

Based on the data in the table, which of the following is closest to the number of mosquitos after a storm with 5.5 inches of rainfall in the city?

A 36 million

B 46 million

C 50 million

D 66 million

QUADRATIC FUNCTIONS AND EQUATIONS

Unit 5
Exponential Functions
and Equations

DOMAIN AND RANGE OF EXPONENTIAL FUNCTIONS

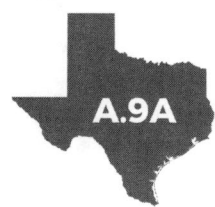

A.9A

The student is expected to determine the domain and range of exponential functions of the form $f(x) = ab^x$ and represent the domain and range using inequalities.

ⓘ TELL ME MORE...

The **domain** of a function is the set of input values that are used for the independent variable. The **range** of a function is the set of output values for the dependent variable. For any exponential function, $f(x) = ab^x$, the domain is the set of all real numbers. The range, however, is bounded by the horizontal asymptote of the graph of $f(x)$. Use the graph to identify the range of $f(x)$ and $g(x)$.

$f(x) = 2^x + 2$
Asymptote: $y = 2$, $a > 0$
Range does not include the asymptote. All real numbers greater than _____.

$g(x) = -(2)^{x-3} - 1$
Asymptote: $y = -1$, $a < 0$
Range does not include the asymptote. All real numbers less than _____.

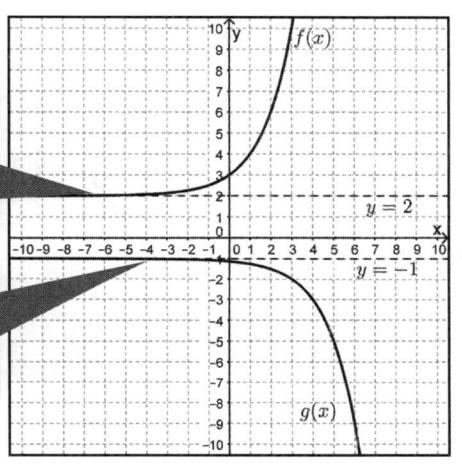

Domain

For any exponential function, $f(x) = ab^x$, the domain is the set of all real numbers.

$$-\infty < x < \infty$$

Range

For any exponential function, $f(x) = ab^x$, the range is the set of real numbers above or below the horizontal asymptote, $y = d$, but does not include d, the value of the asymptote.

$$\text{If } a > 0: f(x) > d$$
$$\text{If } a < 0: f(x) < d$$

However, when an exponential function is used to model real-world situations, then the domain and range can be restricted to numbers that make sense for the situation. In these cases, the domain or range can be written using inequalities for a **continuous** interval.

✓ EXAMPLES

EXAMPLE 1: What are the domain and range of $h(x) = 1.5(3)^x + 4$?

STEP 1 Graph the function to visually inspect the possible range.

- Function values increase as x increases.

- There appears to be a horizontal asymptote near $y = 4$.

- The value of a is positive.

The possible range is $h(x) > 4$.

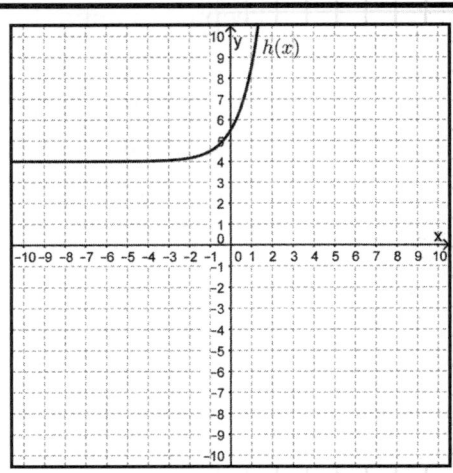

STEP 2 Use the parameters in the equation to confirm the location of the asymptote.

- In $h(x)$, $d = 4$. Thus, the horizontal asymptote is $y = 4$.
- In $h(x)$, $a = 1.5$ which is greater than 0.

The horizontal asymptote is $y = 4$.

STEP 3 State the domain and range of the function.

The domain of $h(x)$ is all real numbers.
The range of $h(x)$ is all real numbers greater than 4, or $h(x) > 4$.

EXAMPLE 2: Charnette purchased a new vehicle. Vehicles depreciate, or lose value over time. Charnette's vehicle depreciates according to the function f whose graph is shown. What is the domain and range of f?

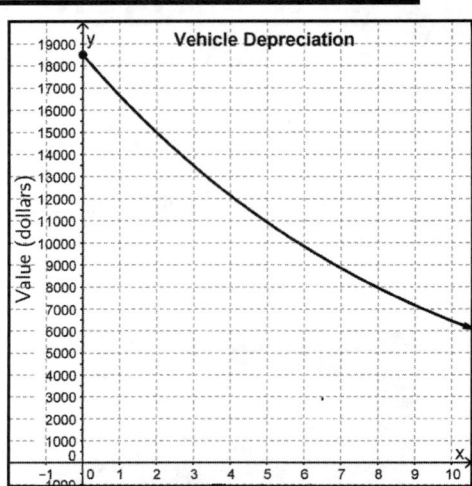

STEP 1 Determine any limits on the independent variable, time, or dependent variable, value.

- In this situation, time is non-negative. It can be 0 or any real number greater than 0, since time is given in years since Charnette purchased the vehicle.
- In this situation, the value of Charnette's car when it was brand new (time = 0) appears to be $18,500. Thus, it cannot exceed this value. The value will also never be less than $0.

Both variables are limited to non-negative numbers (y-axis and Quadrant I) in this situation

STEP 2 State the domain of f for this situation, keeping in mind the constraints already observed.

- x can be any real number greater than or equal to 0.

The domain of the function in this situation is $x \geq 0$.

STEP 3 State the range of the function for this situation, keeping in mind the constraints already observed.

- y, which represents the function values for $f(x)$, can be any real number greater than 0 and less than or including 18,500.

The range of the function in this situation is $0 < y \leq 18{,}500$.

YOU TRY IT!

The table contains some points on the graph of an exponential function. What is the range of the function?

x	y
0	6
1	3
2	1.5
3	0.75
4	0.375

Value of a: _____ b: _____

Function: _____

Range: _____

EXAMPLE 3: The graph of a part of exponential function m is shown. Write the domain and range of the part shown using inequalities.

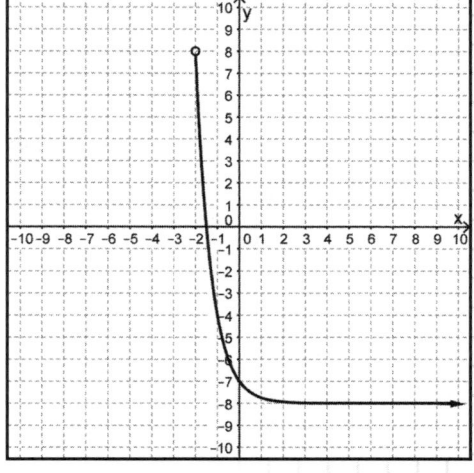

STEP 1 Locate any vertical or horizontal limits such as endpoints or asymptotes.

- open endpoint at $(-2, 8)$
- horizontal asymptote at $y = -8$

STEP 2 Determine any restrictions from the open endpoint.

- There are no x-values to the left of the open endpoint. The lower limit of the domain is -2. Since the endpoint is open, -2 is not included in the domain.

- There are no function values above the endpoint, so the upper limit of the range is 8. Since the endpoint is open, 8 is not included in the range.

–2, not inclusive, is the lower limit of the domain.
8, not inclusive, is the upper limit of the range.

STEP 3 Determine how the asymptote affects the domain and range.

- The horizontal asymptote is the lower limit of the range since the function values will approach, but never equal, the y-value of the asymptote.

- The domain continues to the right without limit.

–8, not inclusive, is the lower limit of the range.

STEP 4 Write the domain using the lower limit, inequalities, and the upper limit. If the limit is included in the domain, be sure the inequality contains "or equal to."

–2 < x

STEP 5 Write the range using the lower limit, inequalities, and the upper limit. If the limit is included in the range, be sure the inequality contains "or equal to."

–8 < y < 8

PRACTICE

Identify the domain and range for each function below using inequalities as appropriate.

1. $f(x) = \frac{1}{2}(3)^x + 2$ when $x \leq 4$

2. $b(x) = 30\left(\frac{1}{2}\right)^x$

3. The table below shows points that belong to the function $g(x)$.

x	-1	0	1	2
$g(x)$	$\frac{1}{3}$	1	3	9

4.

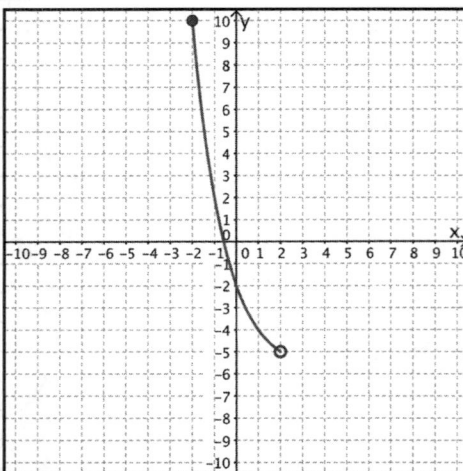

5.

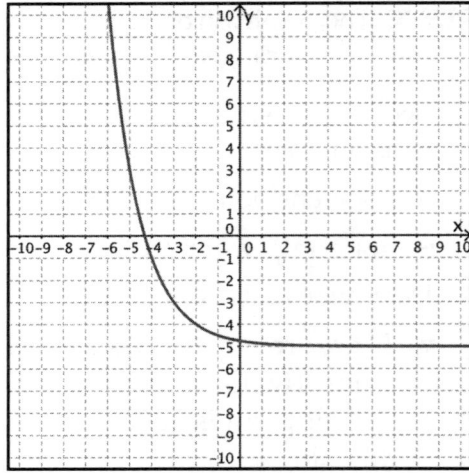

6. The graph below shows a function representing the population of mice, y, in a growing town over a number of years, x. What inequality represents the range of the population of mice in the situation?

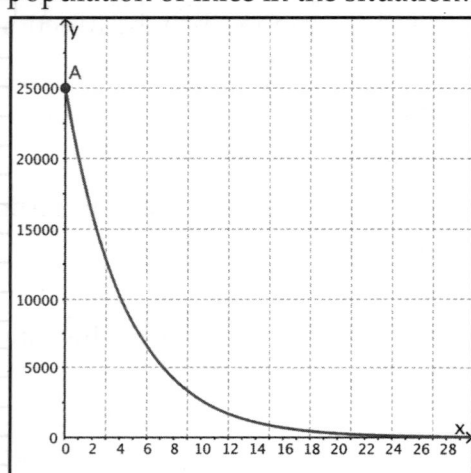

7. The volume, V, of air remaining in an inflated balloon can be modeled by the function $V = 1,000(0.85)^x$ where x represents the number of days that have passed since inflating the balloon. What is the reasonable domain for the situation?

8. The function $f(p) = 65,000(1.05)^x$ can be used to model the population of a city for x, the number years that have passed since 2000. What inequality represents the reasonable range of the function based on the situation?

9. The graph of an exponential function is shown below.

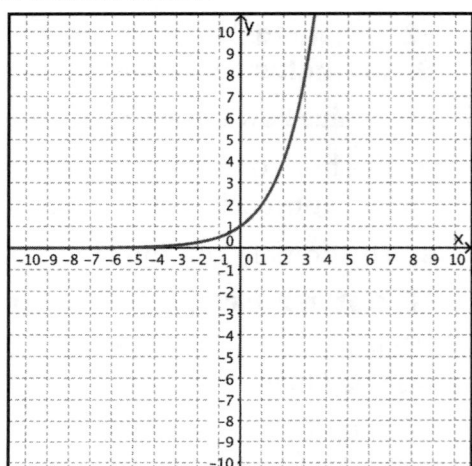

Which of the following statements is true about the function?

A The range of the function is the set of all real numbers.

B The domain of the function is the set of all real numbers less than 4.

C The range of the function is the set of all real numbers less than 4.

D The domain of the function is the set of all real numbers.

INTERPRETING EXPONENTIAL FUNCTIONS

A.9B The student is expected to interpret the meaning of the values of *a* and *b* in exponential functions of the form $f(x) = ab^x$ in real-world problems.

ⓘ TELL ME MORE...

Exponential functions are based on relationships involving a constant multiplier. You can write an exponential function in general form. In this form, a represents an **initial value** or amount, and b, the **constant multiplier**, is a growth factor or factor of decay.

$$f(x) = ab^x$$

a is the initial value or starting point.

b is the growth factor or factor of decay.

Is it Growth or Decay?

The value of *b* answers this question.

- If $b > 1$, then the starting amount, *a*, is multiplied by a number greater than 1 and will increase as *x* increases. This is exponential growth.

- If $0 < b < 1$, then the starting amount, *a*, is multiplied by a number between 0 and 1 and will decrease as *x* increases. This is exponential decay.

✓ EXAMPLES

EXAMPLE 1: Madison is a game warden and monitors the population of fish at a local reservoir. She uses the function $f(x) = 520(2)^x$ to determine the number of fish in the reservoir *x* years after the population was stocked in 2010. What do the numbers 520 and 2 mean in the population model?

STEP 1 Determine the values of *a* and *b* in the function $f(x)$. The coefficient is the value of *a* and the exponential base is the value of *b*.

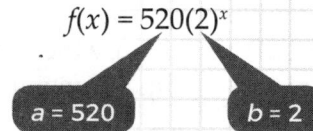
$$f(x) = 520(2)^x$$
a = 520   *b* = 2

a = 520 and b = 2

STEP 2 Interpret the value of a as a starting point or initial value.

- The problem states that $f(x)$ represents the population, or number of fish in the reservoir.

- The problem states that the population of fish was stocked in 2010, meaning that's when the game warden placed fish into the reservoir.

The reservoir was stocked with 520 fish in the year 2010.

STEP 3 Interpret the value of b as the growth rate or rate of decay.

- $b = 2$, which is greater than 1, so this value is a growth rate.

- Multiplying by a factor of 2 results in each year being double the previous year.

The fish population doubles, or increases by a factor of 2, each year.

EXAMPLE 2: Marquise purchased a new motorcycle. The motorcycle depreciates, or loses value, each year. The table contains values of the exponential function $v(x)$. In this function, x represents the number of years since Marquise purchased the motorcycle and $v(x)$ represents the value of the motorcycle. If $v(x) = ab^x$, what do the values of a and b mean in the context of this situation?

Time (years), x	Value (dollars), $v(x)$
0	9000
1	8325
2	7700.63
3	7123.08
4	6588.85

STEP 1 Determine the constant multiplier, or constant ratio, between successive function values. This multiplier will be the value of b.

Time (years), x	Value (dollars), $v(x)$
0	9000
1	8325
2	7700.63
3	7123.08
4	6588.85

$8325 \div 9000 = 0.925$

$7700.63 \div 8325 \approx 0.925$

$7123.08 \div 7700.63 \approx 0.925$

$6588.85 \div 7123.08 \approx 0.925$

$b = 0.925$

MAKE A NOTE . . .

What happens when you multiply a number by 100%, a number less than 100%, or a number greater than 100%? Record your thoughts here.

STEP 2 Interpret the value of b as the growth rate or rate of decay.

- $b = 0.925$, which is less than 1, so this value is a rate of decay and the function values will decrease as x increases.

- 0.925 as a percent is 92.5%. Subtract this value from 100% to determine the annual rate of depreciation.

The value of the motorcycle decreases by 7.5% each year.

STEP 3 Identify the initial value, or the function value when $x = 0$. This value will be the value of a.

$a = 9000$

Time (years), x	Value (dollars), $v(x)$
0	9000
1	8325
2	7700.63
3	7123.08
4	6588.85

STEP 4 Interpret the value of a as a starting point or initial value.

- The problem states that $v(x)$ represents the value of the motorcycle.

- Year 0 was when Marquise purchased the motorcycle.

Marquise purchased the motorcycle for $9000.

EXAMPLE 3: The graph of $f(x)$, an exponential function, is shown. The function $f(x)$ represents the population, in thousands, of a town in West Texas, after oil was discovered in 2010. Let x represent the number of years since 2010. If $f(x)$ is of the form $f(x) = ab^x$, what do the values of a and b mean?

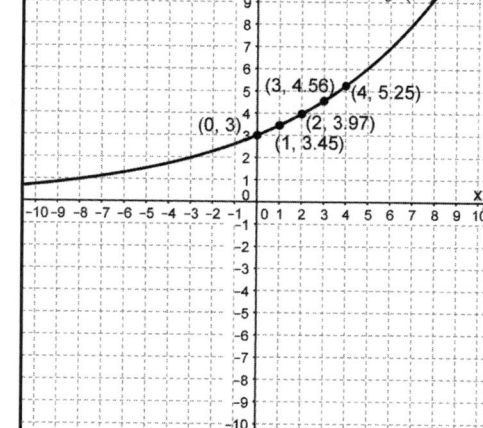

STEP 1   Determine the constant multiplier, or constant ratio, between successive function values. This multiplier will be the value of b. Use any two consecutive values of $f(x)$ from the graph.

$$b = 3.45 \div 3 = 1.15$$

$b = 1.15$

STEP 2   Interpret the value of b as the growth rate or rate of decay.

- $b = 1.15$, which is greater than 1, so this value is a growth rate.

- 1.15 as a percent is 115%. Subtract 100% from this value to determine the annual rate of increase.

The population increases by 15% each year.

STEP 3   Identify the initial value, or the function value when $x = 0$. In a graph, this value is the y-intercept and the y-coordinate of the y-intercept will be the value of a.

The y-intercept is $(0, 3)$, so $a = 3$.

$a = 3$

STEP 4   Interpret the value of a as a starting point or initial value.

- The problem states that $f(x)$ represents the population x years after oil was discovered in 2010.

- Year 0 was when oil was discovered.

The population in 2010 was 3,000.

PRACTICE

Use the following information for Questions 1 and 2.

A bear is given a tranquilizer in order for veterinarians to provide care to the animal. The amount of milligrams of medication remaining in the bear's blood stream based on the number of hours since the injection is modeled with the function $f(x) = 250(0.75)^x$.

1.   What is the meaning of the value 250 in the function model?

2.   What is the meaning of the value 0.75 in the function model?

Use the following information for Questions 3 and 4.

Bacteria is growing exponentially in a laboratory as shown in the table below. The researcher observes the number of bacteria in the sample each hour after starting the sample. She writes a function to model the bacteria's growth in the form of $f(x) = ab^x$, where x represents the time since starting the sample.

Hours, x	0	1	2	3	4	5
Population, $f(x)$	15	30	60	120	240	480

3. What is the value and meaning of the variable a in the function modeling the bacteria's growth?

4. What is the value and meaning of the variable b in the function modeling the bacteria's growth?

5. A rodent species has a population of 25,000 and is decreasing in size at a rate of 15% per year. When the function is written to model the situation, what are the values of a and b used in the function model?

6. The function $f(x) = 300(1.06)^x$ models the value of Mario's investment in a stock fund as a function of time in years. Which of the following is true?

  A Mario initially invested $300 and his money increases by 1.06 times the number of years invested.

  B Mario initially invested $300 and his money decreases by 1.06 times the number of years invested.

  C Mario initially invested $300 and his money increases by 6% each year invested.

  D Mario initially invested $300 and his money decreases by 6% each year invested.

7. The table below shows values for an exponential function, $N(x)$.

Time (years), x	Number in sample, $N(x)$
0	2
1	24
2	288
3	3,456
4	41,472

Which of the following best describes the situation represented by the function?

  F The number of weeds increases by a factor of 12 each year compared to the year before.

  G There were 12 weed plants in the sample when the growth started.

  H The number of weeds increases by 12 times the number of years growing.

  J The number of weeds in the sample is decreasing by a factor of 12 each year compared to the year before.

WRITING EXPONENTIAL FUNCTIONS

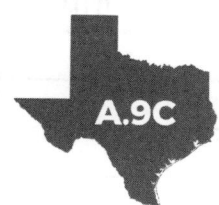

A.9C The student is expected to write exponential functions in the form $f(x) = ab^x$ (where b is a rational number) to describe problems arising from mathematical and real-world situations, including growth and decay.

ⓘ TELL ME MORE...

Exponential functions are based on relationships involving a constant multiplier. You can write an exponential function in general form. In this form, a represents an initial value or amount, and b, the constant multiplier, is a growth factor or factor of decay.

$$f(x) = ab^x$$

Exponential Growth

If $b > 1$, then the starting amount, a, is multiplied by a number greater than 1 and will increase as x increases.

The value of the dependent variable grows as the value of the independent variable increases.

Exponential Decay

If $0 < b < 1$, then the starting amount, a, is multiplied by a number less than 1 and will decrease as x increases.

The value of the dependent variable decays as the value of the independent variable increases.

If you know the starting point, a, and constant multiplier, b, then you can write the exponential function in general form. The values of a and b can be deduced from a table, verbal situation, or diagram. In a table, look for the ratio of successive values of the dependent variable. If these ratios are constant, then the function is an exponential function. The constant ratio is equivalent to the constant multiplier used to generate the exponential function.

✓ EXAMPLES

EXAMPLE 1: The table contains some points on the graph of an exponential function. Based on the table, what function represents the same relationship?

x	y
0	0.04
1	0.2
2	1
3	5
4	25

STEP 1 Determine the constant multiplier, or constant ratio, between successive function values. This multiplier will be the value of b.

x	y
0	0.04
1	0.2
2	1
3	5
4	25

$0.2 \div 0.04 = 5$

$1 \div 0.2 = 5$

$5 \div 1 = 5$

$25 \div 5 = 5$

$b = 5$

STEP 2 Identify the initial value, or the function value when $x = 0$. This value will be the value of a.

a = 0.04

STEP 3 Use the values of a and b that you determined in the table to write the function in $y = ab^x$ form.

a = 0.04 and b = 5, so y = 0.04(5)^x.

x	y
0	0.04
1	0.2
2	1
3	5
4	25

EXAMPLE 2: In January, the Parks and Wildlife Department released 267 bass into a newly constructed pond. Each month, the population of bass in the pond increases by 4.2%. At this rate of growth, what function can be used to determine the population of bass m months after January?

STEP 1 Determine the growth rate or rate of decay. This rate will be the value of b.

Each month, the population increases by 4.2%, so the growth rate is 100% + 4.2%, or 104.2% each year. Percents should be written as their decimal equivalents.

b = 1.042

STEP 2 Identify the initial value, or number of bass at the beginning. This value will be the value of a.

The Parks and Wildlife Department released 267 bass into a pond with no fish. The initial amount of bass is 267.

a = 267

STEP 3 Use the values of a and b that you interpreted from the situation to write the function in $p(m) = ab^m$ form. Let m represent the number of months and $p(m)$ represent the population as a function of m.

$a = 267$ and $b = 1.042$, so

p(m) = 267(1.042)^m.

YOU TRY IT!

The table contains some points on the graph of an exponential function. Based on the table, what function represents the same relationship?

x	y
0	64
1	48
2	36
3	27
4	20.25

Value of a: _____ b: _____

Function: _____

EXAMPLE 3: A pyramid at Chichén Itzá in Mexico contains several layers that are in the shape of square prisms. The first layer has a side length of 50 meters and each successive layer has a side length that is 95% of the one directly below it. What function can be used to find the side length, in meters, of layer L, where $1 \leq L \leq 6$?

El Osario Pyramid, Chichén Itzá, Mexico

STEP 1 Determine the growth rate or rate of decay. This rate will be the value of b.

Each successive layer has a side length that is 95% of the one directly below it, so the rate of decay is 95%. In an equation, percents should be written as their decimal equivalents.

$b = 0.95$

STEP 2 Identify the initial value of the dependent variable, the side length of the first layer. This value will be the value of a.

The ground layer is the first layer. The side length of the ground layer is the initial value, 50 meters.

$a = 50$

MAKE A NOTE . . .

How can you tell if a problem involves exponential growth or exponential decay? Record your thoughts here.

STEP 3 Use the values of a and b that you interpreted from the situation to write the function in $s(L) = ab^L$ form. Let L represent the number of layers above the base layer and $s(L)$ represent the side length as a function of L.

$a = 50$ and $b = 0.95$, so **$s(L) = 50(0.95)^L$**.

PRACTICE

For each table below, determine if the data represent exponential growth or exponential decay. Then, write the function in $f(x) = ab^x$ form.

1.

x	$g(x)$
0	100
1	10
2	1
3	0.01
4	0.01

2.

x	$h(x)$
0	2
1	3
2	$4\frac{1}{2}$
3	$4\frac{3}{4}$
4	$10\frac{1}{8}$

3.

x	$b(x)$
1	20
2	200
3	2000
4	20,000
5	200,000

4.

m	j(m)
0	1.2
1	0.6
2	0.3
3	0.15
4	0.075

5.

t	p(t)
0	210
1	84
2	33.6
3	13.44
4	5.376

6.

r	D(r)
1	6
2	36
3	216
4	1296
5	7776

For each situation below, write an exponential function that can be used to describe the situation.

7. An official NBA basketball must be inflated so that when it is dropped, it must not bounce back more than 75% of the height from which it was dropped. If a basketball is dropped from a height of 72 inches, what function, $r(b)$, describes the height of the basketball as a function of b, the number of bounces?

8. The population of Bexar County, Texas, for a few recent years is shown in the table.

Year	Year Since 2010	Population (millions)
2010	0	1.723
2011	1	1.756
2012	2	1.789
2013	3	1.822
2014	4	1.856

If t represents the number of years since 2010 and $P(t)$ represents the population in millions, what function, $P(t)$ best describes the data in the table?

9. A chiropterologist is a scientist who studies bats. A bat colony is discovered and a chiropterologist calculates that there are 2,100 bats in the colony. If the population of the colony doubles each year, which function, $B(t)$, describes the population of the colony t years after discovery?

A $B(t) = 2100\left(\frac{1}{2}\right)^t$

B $B(t) = 2100(2)^t$

C $B(t) = 2100 + (2)^t$

D $B(t) = 2100 + \left(\frac{1}{2}\right)^t$

10. Chuy purchased a used truck for $11,500. According to an online vehicle website, his truck will depreciate, or lose value, at a rate of 5.5% each year. What function, $d(x)$, represents the value of Chuy's truck x years after its purchase?

F $d(x) = 11,500(0.945)^x$

G $d(x) = 11,500(1.055)^x$

H $d(x) = 11,500(0.055)^x$

J $d(x) = 11,500(5.5)^x$

A.9D The student is expected to graph exponential functions that model growth and decay and identify key features, including *y*-intercept and asymptote, in mathematical and real-world problems.

ⓘ TELL ME MORE...

The graph of an exponential function shows certain key features that are important to the function. For example, the graphs of $f(x) = 2^x$ and $g(x) = -(2)^x$ are shown. Key features of the graphs include the *y*-intercept and horizontal asymptote.

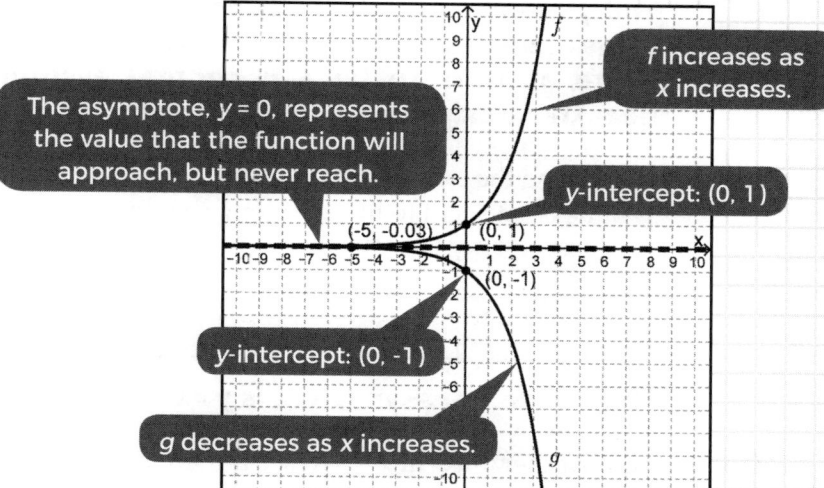

The asymptote, *y* = 0, represents the value that the function will approach, but never reach.

f increases as *x* increases.

y-intercept: (0, 1)

y-intercept: (0, -1)

g decreases as *x* increases.

- The **y-intercept** is the point where the graph of the line crosses the *y*-axis ($x = 0$).

- The **horizontal asymptote** of an exponential function is a horizontal line that represents a particular *y*-value that the exponential function will get very close to but will never be equal to. For the exponential parent function, the horizontal asymptote is the line $y = 0$, which coincides with the *x*-axis.

✓ EXAMPLES

EXAMPLE 1: The graph of an exponential function, *f*, is shown. Identify the *y*-intercept and the line best represents the asymptote of the graph.

STEP 1 Identify the point where the graph of *f* crosses the *y*-axis ($x = 0$).

(0, 3)

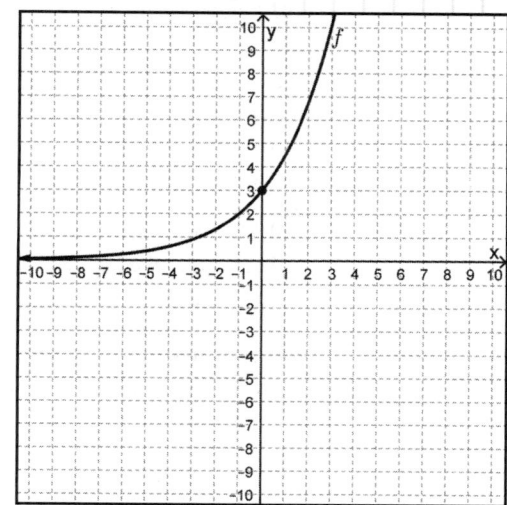

STEP 2 Identify the asymptote as a line near a part of the graph that appears to be flat.

When $x < -7$, the graph of *f* appears to be very close to the line $y = 0$, which is the *x*-axis.

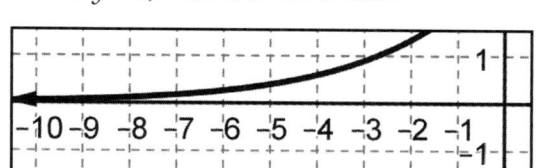

The graph of *f* is close to the line $y = 0$ when $x < -7$. The asymptote is $y = 0$.

EXAMPLE 2: Which of the following statements about the graph of $y = 15\left(\frac{3}{8}\right)^x$ is true?

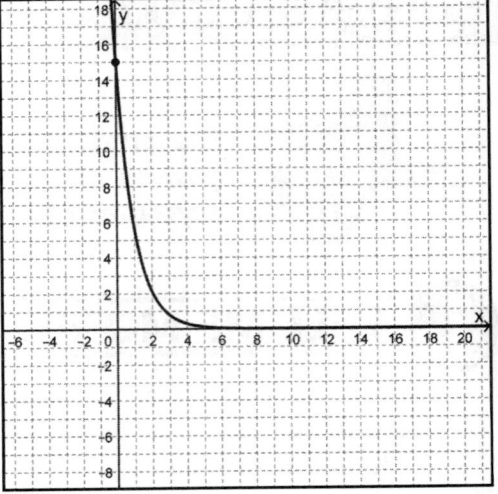

I. The graph has a horizontal asymptote at $y = 0$.

II. The graph has a vertical asymptote at $x = 0$

III. The y-intercept is $(0, 15)$.

IV. The graph decreases from left to right.

STEP 1 Graph y to get an estimate of where the asymptote and y-intercept might be.

The y-intercept appears to be near $(0, 15)$. The value of a in the equation, 15, is the starting point. There appears to be a horizontal asymptote near the x-axis, $y = 0$.

STEP 2 Determine the exact location of the horizontal asymptote. Use a table of values for y around x-values that seem to line up along the asymptote.

x	4	5	6	7	8	9
y	0.297	0.111	0.042	0.016	0.006	0.002

The horizontal asymptote is $y = 0$, so Statement I is correct.

STEP 3 Look at the graph at $x = 0$ to see if there is a vertical asymptote at this point.

When $x = 0$, $y = 15$, so there cannot be an asymptote at $x = 0$.

Statement II is not correct.

STEP 4 Locate the y-intercept on the graph. Use the function to confirm its location $(0, y)$.

$$y = 15\left(\frac{3}{8}\right)^x$$

$$y = 15\left(\frac{3}{8}\right)^0$$

$$y = 15(1)$$

$$y = 15$$

The y-intercept is $(0, 15)$. Statement III is correct.

STEP 5 Determine if the function values increase or decrease as x increases from left to right.

■ As x increases from left to right, function values that begin around $(0, 15)$ quickly decrease toward $y = 0$. After $x = 5$, function values continue to slowly decrease as they gradually approach $y = 0$, the horizontal asymptote.

The graph decreases as x increases from left to right. Statement IV is correct.

EXAMPLE 3: The number of views of a popular online video can be modeled by the exponential function graphed on the grid, where x is the number of months since the video was posted. How many views did the video immediately receive when it was posted?

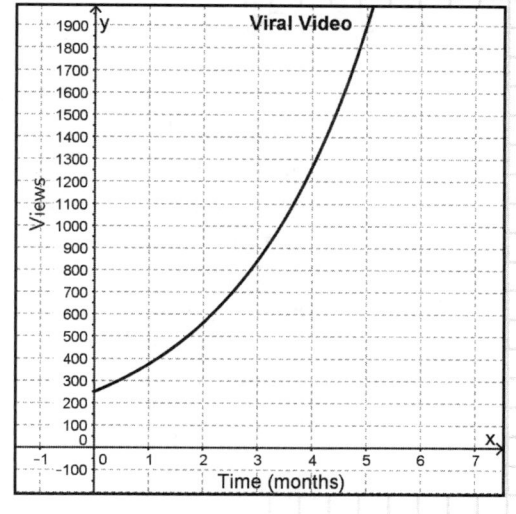

STEP 1 The independent variable, time since posting, is 0 which represents when the video was immediately posted. Determine the y-intercept.

From the graph, when $x = 0$, $y \approx 250$.

(0, 250) is the y-intercept.

STEP 2 Interpret the y-intercept to answer the question from the problem.

How many views did the video immediately receive when it was posted?

The video was posted at time = 0, or $x = 0$. The value of the dependent variable, the number of views, tells you the number of views when $x = 0$, or when the video was immediately posted. This number is the y-coordinate of the y-intercept.

250 views

PRACTICE

1. The graph of exponential function $k(x)$ is shown below.

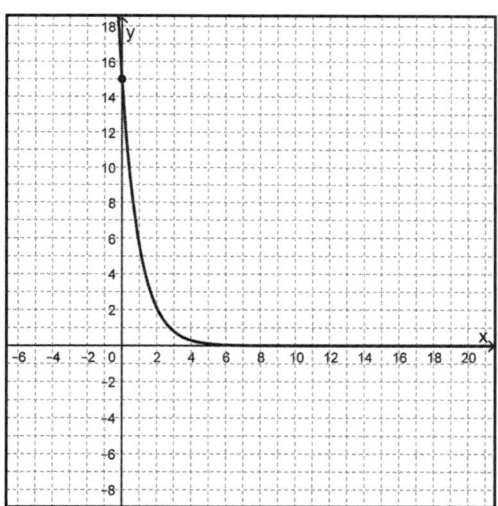

What is the equation of the line that forms an asymptote for the graph of the function?

2. What are the x- and y-intercept values of $f(x) = -2(0.5)^x + 8$?

3. For the graph of exponential function $g(x) = 500(0.85)^x$, describe the behavior of the graph as it moves left to right. What is its y-intercept and the equation for any asymptote of the graph?

4. A motorcycle valued at $6,500 depreciates at a rate of 12% annually. What is the y-intercept of the graph of the function that models the situation and what is the equation of any asymptote to the function's graph?

5. A biologist is studying a new species of bacteria. The biologist starts a sample of 100 bacteria and observes the growth. The function $g(x) = 100(2)^x$ models the growth of the bacteria where x represents the number of 6-hour time periods and $g(x)$ represents the number of bacteria in the culture sample. What is the shape of the graph of the function and what is the y-intercept of the graph?

6. A scientist is studying the number of mold cells present on food items based on the time it is allowed to grow and reproduce. The function $m(x) = 50(3)^x$ represents the number of mold cells present on the food for each observation time interval, x. What is the asymptote of the graph of the function? What is the y-intercept?

7. LaToya received $100 from her grandmother to open a savings account for her birthday. The account LaToya opens pays an annual interest rate of 4%. The function $y = 100(1.04)^x$ represents LaToya's bank balance, y, given the time of years, x, the account has been open. What is the y-intercept of the graph of the function? How does the graph behave as it moves from left to right?

8. The functions $f(x) = \left(\frac{1}{2}\right)^x$ and $g(x) = 2^x$ are plotted together on the same graph. What is the difference in the asymptotes of the two graphs?

9. For the exponential function shown graphed below, which is an equation for the asymptote of the function?

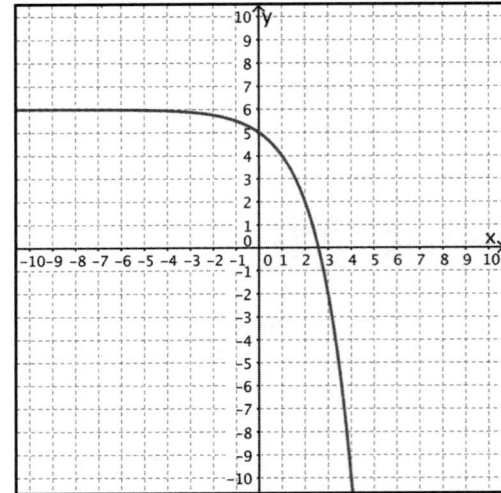

A $x = 4$

B $y = 6$

C $x = 6$

D $y = 4$

10. Which of the following is not true of the graph of $f(x) = \frac{5}{2}(3)^x$?

F There is an asymptote at $y = 0$.

G The graph increases from left to right.

H There is an asymptote at $x = 2$.

J The y-intercept is $(0, 2.5)$.

11. Reginald received a job offer from a local company with a starting annual salary of $30,000. Each year, he will receive an annual salary increase of 10%. Which graph models this situation after Reginald receives x annual increases?

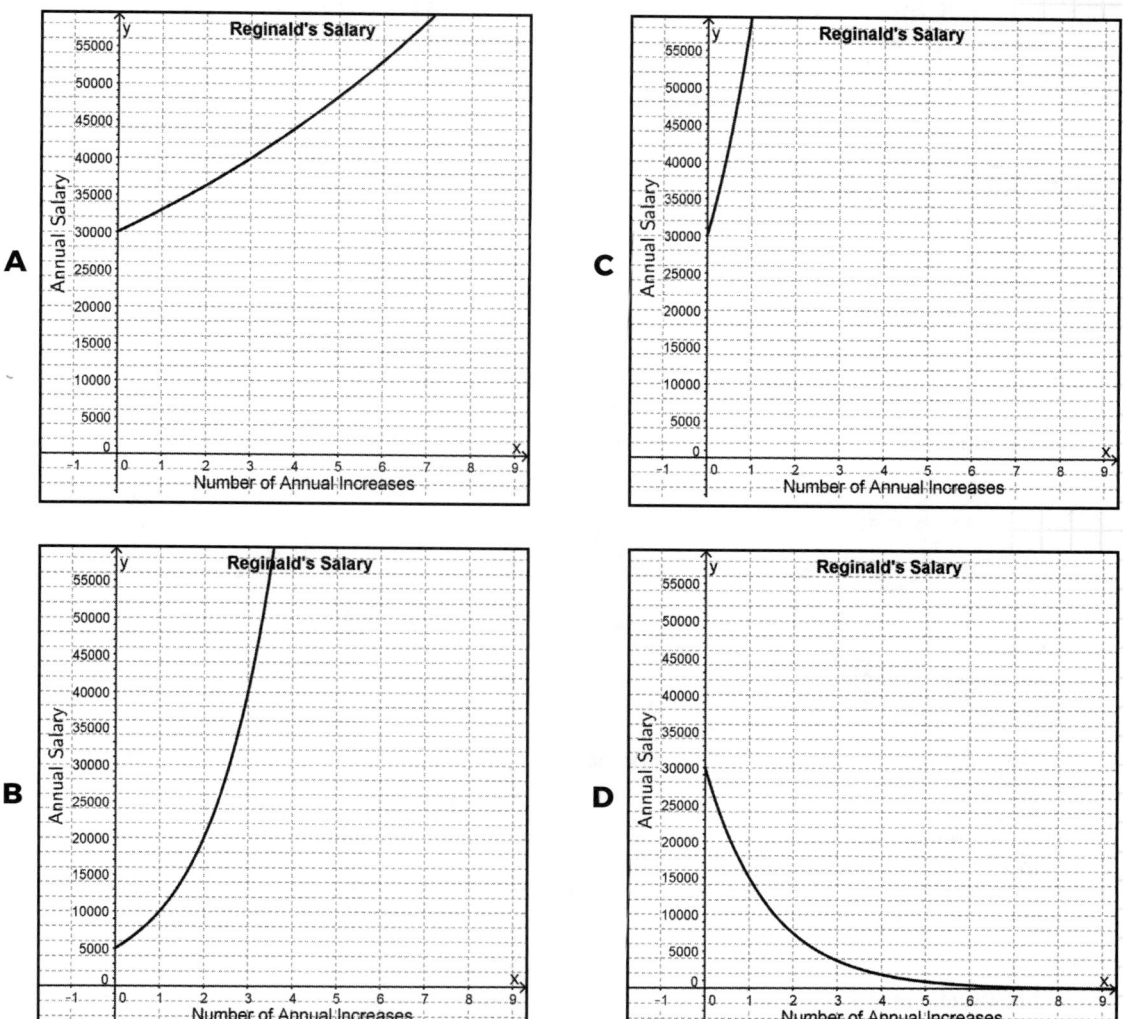

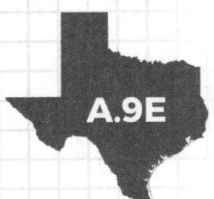

A.9E   The student is expected to write, using technology, exponential functions that provide a reasonable fit to data and make predictions for real-world problems.

ⓘ TELL ME MORE...

Exponential functions are based on a constant multiplier, or base, that is multiplied by itself x times. Repeated multiplication is represented with an exponent. Hence, the independent variable in an exponential function appears in the exponent of the function. You can write an exponential function in general form, $f(x) = ab^x$, where a and b are real numbers, b represents the constant multiplier, or base, and a represents an initial value or starting point.

$$f(x) = ab^x$$

Modeling with Transformations

Make a scatterplot of the data. Beginning with the parent function $f(x) = 2^x$, adjust the starting point, a, to match your data. Then, adjust the base, b, so that the curve aligns with the data points.

Modeling with Regression

Enter the data into a spreadsheet, graphing technology, or app. Use the technology's regression programming to generate a curve of best fit. Graph the curve on top of a scatterplot of your data to make sure the curve fits.

✓ EXAMPLES

EXAMPLE 1: LaQuan is a medical technologist studying bacteria cultures. For one particular bacteria strain, he started with 200 bacteria in a Petri dish. The table shows the growth of the bacteria culture over time. Based on the data, write a function that best models the data and use the function to predict when the bacteria culture will reach 4000 bacteria.

Time (days), x	Number of Bacteria, $p(x)$
0	200
1	280
2	390
3	550
4	768
5	1075
6	1500

STEP 1   Make a scatterplot of the data. Graph the exponential parent function, $f(x) = 2^x$, over the scatterplot.

$f(x) = (2)^x$

Number of Bacteria

Time (days)

STEP 2 Transform the value of *a*, the starting point, so that it aligns with (0, 200), the starting point from the table.

- In the general form $f(x) = ab^x$, let a equal the value of the starting point, or a = 200.

- Graph $f(x) = 200(2)^x$

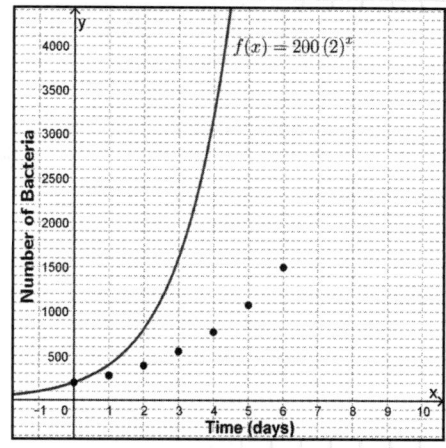

STEP 3 Transform the curvature of the graph by adjusting the base, *b*.

- Use the data in the table to estimate a value of *b*. Look for a constant ratio between successive function values.

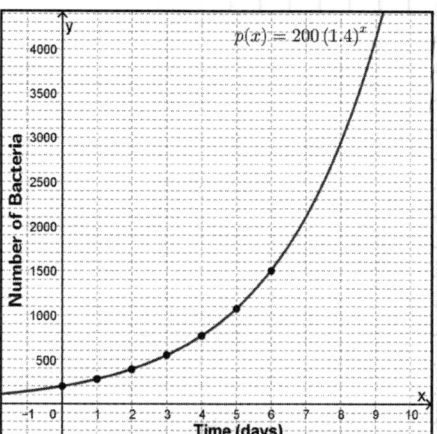

Time (days), *x*	Number of Bacteria, *p(x)*
0	200
1	280
2	390
3	550
4	768
5	1075
6	1500

$280 \div 200 = 1.4$

$390 \div 280 \approx 1.39$

$550 \div 390 \approx 1.41$

$768 \div 550 \approx 1.40$

- For the first few rows of data, you multiply the previous function value by approximately 1.4 to generate the next function value. Graph $p(x) = 200(1.4)^x$.

$p(x) = 200(1.4)^x$

STEP 4 The number of bacteria is the dependent variable. Let $p(x) = 4000$ and graphically estimate the solution to the equation $4000 = 200(1.4)^x$.

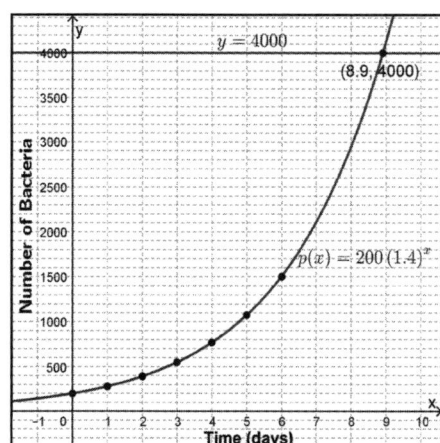

The number of bacteria will reach 4000 in about 8.9 days.

<table>
<tr><td colspan="2" align="center">YOU TRY IT!</td></tr>
</table>

Write a function that models the data shown.

x	*y*
0	75
1	262
2	920
3	3200
4	11,225

Value of *a*: _____ *b*: _____

Function: _____

EXAMPLE 2: A company that manufactures car parts found a new process that allows them to reduce the cost to manufacture a certain part. The table shows the cost per part for the first few months of the process. Write $c(x)$ that describes the cost as a function of time.

Month, x	Cost per Part, $c(x)$
0	600
1	480
2	380
3	310
4	250
5	200
6	155
7	125
8	100

STEP 1 Use regression with technology to determine an appropriate function model. Enter the data into your technology.

L1	L2	L3	L4	L5
0	600	------	------	------
1	480			
2	380			
3	310			
4	250			
5	200			
6	155			
7	125			
8	100			
------	------			

STEP 2 Use the technology's exponential regression algorithm to generate an exponential regression equation.

$c(x) = 602.197(0.799)^x$

```
          ExpReg
y=a*b^x
a=602.197417
b=0.7993790531
r²=0.9996543862
r= -0.9998271782
```

STEP 3 Graph the exponential regression equation for $c(x)$ over a scatterplot of the data. Compare the graph of the function model with the scatterplot of the actual data.

The graph seems to pass through or very near every data point in the scatterplot. The graph follows the general trend of the data.

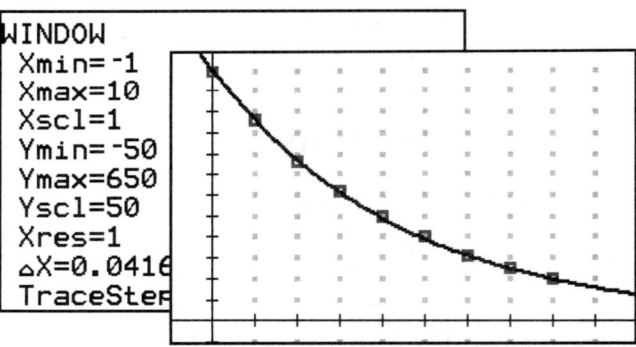

```
WINDOW
 Xmin= -1
 Xmax=10
 Xscl=1
 Ymin= -50
 Ymax=650
 Yscl=50
 Xres=1
 ΔX=0.0416
 TraceStep
```

EXAMPLE 3: The median sale price of a single family home in a city in 2010 was $155,000. The table shows how that value changed over the next several years. If prices continue changing at this rate, write an exponential function, $f(x)$, that best describes the data. What will be the median sale price of a single family home in 2020?

Year since 2010, x	0	1	2	3	4
Sale Price, $f(x)$	$155,000	$164,300	$174,200	$185,000	$192,000

STEP 1 Use regression with technology to determine an appropriate function model. Enter the data into your technology.

L1	L2	L3	L4	L5
0	155000	------	------	------
1	164300			
2	174200			
3	185000			
4	192000			
------	------			

STEP 2 Use the technology's exponential regression algorithm to generate an exponential regression equation.

f(x) = 155597.79(1.056)ˣ

ExpReg
y=a*b^x
a=155597.7916
b=1.05620282
r²=0.9939626735
r=0.9969767668

STEP 3 Determine the function value for 2020. In this case, the input value (independent variable) is the year since 2010, x, which is 2020 – 2010 = 10. Substitute $x = 10$ into $f(x)$ and evaluate the function.

$$f(x) = 155597.79(1.056)^x$$
$$f(10) = 155597.79(1.056)^{10}$$
$$f(10) \approx 155597.79(1.7244)$$
$$f(10) \approx 268313$$

The median sale price in 2020 will be approximately $268,313.

PRACTICE

1. A table of values for the exponential function g is shown. What is the function that is best represented by the data in the table?

x	g(x)
0	250
1	125
2	62.5
3	31.25
4	15.625

2. The first 5 figures of a pattern are shown below. Each figure of made up of identical cubes. If the pattern continues, what expression can be used to find the number of cubes that make up Figure n?

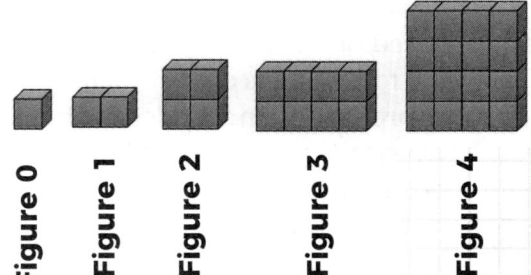

Figure 0   Figure 1   Figure 2   Figure 3   Figure 4

3. Araceli's grandmother opened a savings account for her 10 years ago. The account carries an annual interest rate. The table below shows the balance in the account for some of the years since it was opened.

Year, x	Balance (dollars), y
1	$1,590.00
3	$1,786.50
6	$2,217.80
8	$2,390.80
10	$2,686.30

Based on the data, to the nearest percent, what is the approximate interest rate?

4. The graph shows data for the mass of a sample of a radioactive substance based on the time since it began to decay.

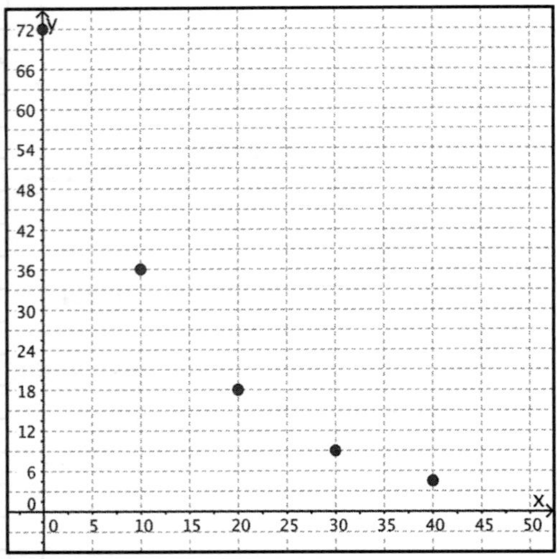

Based on the data, which of the following is closest to the approximate time of decay after which the substance will have a mass of 1 gram?

A 50 hours

B 62 hours

C 67 hours

D 100 hours

5. The table shows the temperature of a cup of coffee at various times after being poured into a cup and left to cool.

Time (minutes), x	Temperature (°F), $f(x)$
0	179.5
5	168.7
10	150.3
15	141.7
20	130.0

Based on the data in the table, after about how many minutes will the approximate temperature of the cup of coffee reach 100°F?

6. Carla's pet puppy received an injection. The table shows the amount of medication, M, remaining in the puppy's bloodstream over a period of time, h.

Time (hr)	Medication (mg)
0	275
1	220
2	176
3	140.8
4	112.64
5	90.112

Based on the information in the table, what function rule best models the relationship between the number of hours and the number of milligrams of the medication remaining in the puppy's bloodstream?

F $M = 0.8^h + 275$

G $M = 0.8 \cdot 275^h$

H $M = 0.8h + 275$

J $M = 275 \cdot 0.8^h$

PERSONAL GLOSSARY

Term	Picture and Definition

PERSONAL GLOSSARY

Term	Picture and Definition

PERSONAL GLOSSARY

Term	Picture and Definition

Term	Picture and Definition

PERSONAL GLOSSARY

Term	Picture and Definition

PERSONAL GLOSSARY

Term	Picture and Definition

PERSONAL GLOSSARY

Term	Picture and Definition

PERSONAL GLOSSARY

Term	Picture and Definition